*Practical Fluid Mechanics
for Engineers and Scientists*

HOW TO ORDER THIS BOOK

BY PHONE: 800-233-9936 or 717-291-5609, 8AM–5PM Eastern Time

BY FAX: 717-295-4538

BY MAIL: Order Department
Technomic Publishing Company, Inc.
851 New Holland Avenue, Box 3535
Lancaster, PA 17604, U.S.A.

BY CREDIT CARD: American Express, VISA, MasterCard

PRACTICAL FLUID MECHANICS FOR ENGINEERS AND SCIENTISTS

NICHOLAS P. CHEREMISINOFF

TECHNOMIC
PUBLISHING CO., INC.
LANCASTER · BASEL

Practical Fluid Mechanics for Engineers and Scientists
a TECHNOMIC®publication

Published in the Western Hemisphere by
Technomic Publishing Company, Inc.
851 New Holland Avenue
Box 3535
Lancaster, Pennsylvania 17604 U.S.A.

Distributed in the Rest of the World by
Technomic Publishing AG

Copyright © 1990 by Technomic Publishing Company, Inc.
All rights reserved

No part of this publication may be reproduced, stored in a
retrieval system, or transmitted, in any form or by any means,
electronic, mechanical, photocopying, recording, or otherwise,
without the prior written permission of the publisher.

Printed in the United States of America
10 9 8 7 6 5 4 3 2 1

Main entry under title:
 Practical Fluid Mechanics for Engineers and Scientists

A Technomic Publishing Company book
Bibliography: p.
Index p. 271

Library of Congress Card No. 90-70985
ISBN No. 87762-639-1

TABLE OF CONTENTS

PREFACE **ix**

ABOUT THE AUTHOR **xi**

CHAPTER 1 **Elements of Fluid Mechanics** **1**
    Units of Measure   **1**
        Prefixes for Decimal Multiples and Submultiples of SI Units   **5**
    Industrial Stoichiometry   **6**
        Fundamental Quantities of Matter   **6**
        Internal Energy and the First Law of Thermodynamics   **14**
        Enthalpy   **16**
        Heat Capacity and Specific Heat   **16**
        Properties of the Ideal Gas   **17**
    Behavior of Fluids   **21**
        Time-Independent Fluids   **25**
        Time-Dependent Fluids   **31**
        Viscoelastic Fluids   **34**
    Types of Operations and Approaches to Problem Analysis   **35**
    Notation   **40**
        Greek Symbols   **40**
    References   **41**

CHAPTER 2 **Similarity, Modeling, and Dimensional Analysis** **43**
    Introduction   **43**
    Principles of Similarity Theory   **44**
    Methods of Similarity Theory   **46**
        Newton's Theorem   **47**
        Derivation of Dimensionless Groups from Process-Governing Equations   **52**
        Dimensional Analysis   **56**
        Buckingham Pi Theorem   **58**

Federman-Buckingham's Theorem 59
Kirpichev-Gukhman's Theorem 61
Principles and Methods of Modeling Unit Operations 62
Notation 69
Greek Symbols 69
References 70

CHAPTER 3 **Hydraulic Processes** 71

Introduction 71
Basic Definitions 71
Physical Properties of Fluids 73
Density and Specific Gravity 73
Viscosity 75
Pressure and Surface Tension 81
Liquid Pressure 81
Surface Tension 82
Notation 83
Greek Symbols 84
References 84

CHAPTER 4 **Hydrostatics** 85

Introduction 85
Euler's Differential Equations 85
The Basic Equation of Hydrostatics 87
Application of Hydrostatic Principles to Manometric
 Techniques 90
Pressure Forces Acting on Submerged Flat Surfaces 101
Hydrostatic Machines 111
Notation 112
Greek Symbols 112
References 113
Bibliography 113

CHAPTER 5 **Hydrodynamics: Single-Fluid Flows** 115

Introduction 115
Characteristics of Fluids in Motion 115
Steady-State versus Transient Flows 115
Fluid Mass and Linear Velocities 117
Residence Time 121
Regimes of Flow 124
Laminar versus Turbulent Flows 124
Turbulence 130
Continuity Equation 135
Equations of Motion for Ideal Fluids 137

Differential Equations for Viscous Fluids   139
   Navier-Stokes Equations   139
   Transformation Techniques   145
The Total Energy Balance   148
   Discharge of Gases   151
   Gas Flow through Piping   156
The Bernoulli Equation   170
   General Application   170
   Variable-Head Meters   178
   Efflux from Vessels and Pipes   187
Hydraulic Resistances in Pipe Flow   194
   Flow through Pipes   194
   Flow through Varying Cross Sections   208
   Flow through Piping Components   218
   Velocity Dampening   233
Flow Normal to Tube Banks   236
Optimum Pipe Diameter   239
Flow of Liquid Films   247
   Vertical Film Flow   247
   Wavy Flow   257
   Turbulent Film Flow   262
Notation   265
   Greek Symbols   267
References   268
Bibliography   269

INDEX   **271**

PREFACE

This book is intended to provide a convenient summary of fundamentals for engineers, scientists, technicians, plant operations personnel. A handy reference on fluid flow, it is the key to numerous unit operations in the process industries as well as in many pollution control and environmental processes and materials handling.

Principles and methods are presented for application in fluid transport. Basic principles of fluid mechanics have been related to design and for equipment design and selection in terms of operational fundamentals. It is intended as a quick and ready reference.

NICHOLAS P. CHEREMISINOFF

ABOUT THE AUTHOR

Nicholas P. Cheremisinoff heads the Product Development Group of the Elastomers Technology Division of Exxon Chemical Company, Linden, New Jersey. He is involved in product research and development of polymeric materials for the consumer market and design/development of multiphase reactors and instrumentation for complex process control. He is the author/co-author of numerous books and papers on the subject of applied fluid mechanics. Dr. Cheremisinoff received his B.S., M.S. and Ph.D. degrees in Chemical Engineering from Clarkson College of Technology. He is a member of several professional and honor societies, including Tau Beta Pi and Sigma Xi.

CHAPTER 1
Elements of Fluid Mechanics

UNITS OF MEASURE

First, we will define an engineering system of units that provides a standard format for measuring, interpreting, and reporting phenomenological evidence, as well as a vehicle by which such information may be applied to benefit mankind. Historically, units of measure have been arbitrary; investigators working in a particular field would elect to devise convenient definitions. To standardize definitions of measure, the scientific and engineering communities have formulated a policy of an International System of Units (SI) through the International Organization of Standardization (ISO).

Transition to SI has advanced considerably during the past decade. The Metric Conversion Act of 1975 (PL 94-168) declared the coordination and planning of increasing use of the metric system (SI) in United States government policy. A memorandum by the Assistant Secretary of Commerce for Science and Technology in the *Federal Register* of October 26, 1977 interprets and modifies SI for the United States. The Act also provides for the establishment of the U.S. Metric Board to coordinate voluntary conversion. Despite this increased activity, complete changeover to SI has not yet materialized. Consequently, in this transition phase, practicing engineers, scientists, and students must be familiar with several systems of units: English–British units, cgs, and SI. This book applies all three systems to problem solving, with emphasis on SI.

As noted, SI is the abbreviation for the International System of Units (Le Système International d'Unités). The prelude to SI was the cgs (centimeter-gram-second) system of metric units. In contrast, SI is based on the meter, kilogram and second as the fundamental units.

The SI system utilizes three classes of units. The first comprises *base units*. By convention, these are dimensionally independent. The second includes *supplementary units,* which are used to measure plane and solid angles. *Derived units* are those formed by algebraic combinations of base units, supplementary units, and other derived units. Specific names and symbols are assigned to the units in each class.

The advantage of SI over English units is the elimination of conversion factors within the system. That is, all derived combinations are in terms of unity. For example, the derived unit of power is the *watt,* which in base units, is defined as 1 joule of work completed in 1 second of time.

There are seven *base units,* each considered dimensionally independent with specific definitions:

- meter (m) for length
- kilogram (kg) for mass
- second (s) for time
- ampere (A) for electric current

Table 1.1. SI Base Units and Their Definitions
(courtesy of the American Petroleum Institute — API Pub. 2564).

Quantity	Name	Symbol	Definition
Length	meter (or metre)	m	The meter is the length equal to 1 650 763.73 wavelengths in vacuum of the radiation corresponding to the transition between the levels $2p_{10}$ and $5d_5$ of the krypton-86 atom (Eleventh CGPM, 1960, Resolution 6).
Mass	kilogram	kg	The kilogram is the unit of mass (not force); it is equal to the mass of the international prototype of the kilogram (Third CGPM, 1901, Resolution 3).
			This international prototype, made of platinum-iridium, is kept at the International Bureau of Weights and Measures. A copy of the international prototype is maintained by the national standards agency of each major country.
			The kilogram is the only base unit defined by an artifact and is the only base unit having a prefix.
Time	second	s	The second is the duration of 9 192 631 770 periods of the radiation corresponding to the transition between the two hyperfine levels of the ground state of the cesium-133 atom (Thirteenth CGPM, 1967, Resolution 1).
Electric Current	ampere	A	The ampere is that constant current that, if maintained in two straight parallel conductors of infinite length of negligible circular cross section and placed 1 meter apart in vacuum, would produce between these conductors a force equal to 2×10^{-7} newton per meter of length (CIPM, 1946, Resolution 2 approved by the Ninth CGPM, 1948).

(continued)

- kelvin (K) for temperature
- mole (mol) for the amount of a substance
- candela (cd) for luminous intensity

Table 1.1 provides the exact definitions of each of these units.

At present, there are only two *supplementary units* in the SI system. Both are purely geometric:

- radian (rad) for the unit of plane angle
- steradian (sr) for the unit of solid angle

Table 1.1. (continued).

Quantity	Name	Symbol	Definition
Temperature	kelvin	K	The kelvin, unit of thermodynamic temperature, is the fraction 1/273.16 of the thermodynamic temperature of the triple point of water (Thirteenth CGPM, 1967, Resolution 4).
			The unit kelvin and its symbol, K, are used to express an interval or difference of temperature (Thirteenth CGPM, 1967, Resolution 3).
			In addition to the thermodynamic temperature, Celsius temperature (formerly called Centigrade) is used widely. The degree Celsius (°C), a derived unit, is the unit for expressing Celsius temperatures and temperature intervals. Celsius temperature, t, is related to thermodynamic temperature, T, by the following equation: $$t = T - T_0$$ where $T_0 = 273.16$ by definition (temperature interval 1°C equals 1 K exactly)
Amount of Substance	mole	mol	The mole is the amount of substance of a system that contains as many elementary entities as there are atoms in 0.012 kilogram of carbon-12.
			When the mole is used, the elementary entities must be specified and may be atoms, molecules, ions, electrons, other particles or specified groups of such particles (Fourteenth CGPM, 1971, Resolution 3).
Luminous	candela	cd	The candela is the luminous intensity, in a given direction, of a source that emits monochromatic radiation of frequency 540×10^{12} hertz and that has a radian intensity in that direction of 1/683 watt per steradian (Sixteenth CGPM, 1979).

Table 1.2. Examples of SI-Derived Units.

Quantity	SI Unit			
	Name	Symbol	Expression in Terms of Other Units	Expression in Terms of SI Base Units
Frequency	hertz	Hz		s^{-1}
Force	newton	N		$m \cdot kg \cdot s^{-2}$
Pressure	pascal	Pa	N/m^2	$m^{-1} \cdot kg \cdot s^{-2}$
Energy, Work, Quantity of Heat	joule	J	N/m	$m^2 \cdot kg \cdot s^{-2}$
Power, Radiant Flux	watt	W	J/s	$m^2 \cdot kg \cdot s^{-3}$
Electric Potential, Potential Differences, Electromotive Force	volt	V	W/A	$m^2 \cdot kg \cdot s^{-3} \cdot A^{-1}$
Electric Resistance	ohm	Ω	V/A	$m^2 \cdot kg \cdot s^{-3} \cdot A^{-2}$
Conductance	siemens	S	A/V	$m^{-2} \cdot kg^{-1} \cdot s^3 \cdot A^2$
Area	square meter	m^2		
Volume	cubic meter	m^3		
Speed, Velocity	meter per second	m/s		
Acceleration	meter per second squared	m/s^2		
Density	kilogram per cubic meter	kg/m^3		
Concentration (of amount of substance)	mole per cubic meter	mol/m^3		
Specific Volume	cubic meter per kilogram	m^3/kg		
Luminance	candela per square meter	cd/m^2		

The radian is the plane angle between two radii of a circle that cut off, on the circumference, an arc equal in length to the radius. The steradian is the solid angle that, having its apex in the center of a sphere, cuts out an area of the surface of the sphere that equals that of a square, with sides of length equal to the radius of the sphere.

Derived units are expressed algebraically in terms of base units, having mathematical symbols for multiplication and division. A number of derived units have been given special names and assigned symbols. Examples of derived units and those given special names are noted in Table 1.2.

Several other units are used widely but are not part of SI. These include the minute, hour, day, and year as units of time; degree, minute and second of arc (in addition to the radian); the metric ton (1000 kg); the liter (1 cubic decimeter); the nautical mile; and the knot. These may be used along with SI units.

Prefixes for Decimal Multiples and Submultiples of SI Units

These are listed in Table 1.3. The symbol of a prefix is combined with the unit to which it is directly attached, thus forming a new unit symbol that can be raised to a positive or negative power and can be combined with other unit symbols to form compound units.

Distinction between upper case and lower case symbols is important, as shown by the following examples:

- M = mega = 10^6
- m = milli = 10^{-3} (where m is a prefix)

Table 1.3. SI Prefixes for Forming Decimal Multiples and Submultiples.

Prefix	Symbol	Factor by Which Unit Is Multiplied
tera	T	10^{12}
giga	G	10^9
mega	M	10^6
kilo	k	10^3
hecto	h	10^2
deca	da	10
deci	d	10^{-1}
centi	c	10^{-2}
milli	m	10^{-3}
micro	μ	10^{-6}
nano	n	10^{-9}
pico	p	10^{-12}
femto	f	10^{-15}
atto	a	10^{-18}

- N = newton
- n = nano = 10^{-9}

Caution must be exercised when a compound unit includes a unit symbol that is also a symbol for a prefix. As an example, the unit newton-meter should be written as N-m to avoid confusion with mN (millinewton).

Symbols can be used together with modifying subscripts and/or superscripts. Subscripts are often used to designate a place in space or time, or a constant or reference point. Superscripts can be used to designate a dimensionless form, reference or equilibrium value, or mathematical identification, such as an average value, derivative, tensor index, etc. Table 1.4 provides a partial list of commonly used symbols and their definitions by category.

INDUSTRIAL STOICHIOMETRY

Fundamental Quantities of Matter

The properties of the three states of matter (solids, liquids, gases) are detected based on quantities related to our senses. These quantities include time, distance, mass, force, and temperature. The first three are obvious. The concept of force is best introduced by way of Newton's second law of motion. This fundamental relation expresses the quantity force in terms of the product of mass and acceleration, or

$$F_0 = \frac{1}{g_c} ma \qquad (1.1)$$

This equation provides a relationship among the four fundamental quantities—time, distance, mass and force—which is valid regardless of the system of units employed. The quantity g_c is a conversion factor based on the appropriate unit of measure. For example, when time is measured in seconds, distance in centimeters and mass in grams, the unit of force is the dyne (i.e., the force that will cause a mass of 1 gram to accelerate 1 cm/s²). Substituting these units into Equation (1.1) gives

$$1 \text{ dyne} = \frac{1}{g_c} (1 \text{ g})(1 \text{ cm/s}^2)$$

or

$$g_c = 1 (\text{g-cm})/(\text{dyne-s}^2)$$

The dyne may be regarded merely as an abbreviation for the composite unit g-cm/s² and, hence, g_c is unity and dimensionless.

In the English engineering system, a pound force is the force exerted by gravity on a 1-lb mass of material under conditions in which the acceleration of gravity is 32.1740 ft/s². Hence, a 1-lb mass will acquire an acceleration of 32.1740 ft/s² by a force of 1 lb-force, or

$$1 \text{ lb}_f = \frac{1}{g_c} (1 \text{ lb}_m)(32.1740 \text{ ft/s}^2)$$

or

$$g_c = 32.1740 \frac{\text{lb}_m\text{-ft}}{\text{lb}_f\text{-s}^2}$$

Thus, in the English system, g_c is a dimensional constant having a numerical value equivalent to the standard of the acceleration of gravity, but with different units.

Temperature is a measure of the degree of hotness of a substance and is detected most commonly with a liquid-in-glass thermometer. By employing a uniform tube partially filled with an appropriate fluid such as mercury, alcohol, or an oil, the property of thermal expansion of the fluid provides a measure when heated. That is, the degree of hotness is detected by measuring the length of the fluid in the column. The numerical values are again arbitrary in origin. In the centigrade or Celsius scale, the freezing point of water saturated with air at standard atmospheric pressure is defined as zero, and the boiling point of pure water at standard atmospheric pressure is assigned the value of 100. Other temperature scales are defined in terms of the Celsius scale. The Fahrenheit scale is defined as follows:

$$t°F = 1.8(t°C) + 32 \qquad (1.2)$$

The freezing point on this scale is 32°F, and the boiling point is 212°F.

Two other temperature scales of primary importance introduce the concept of an absolute lower limit of temperature. Accurate measurements establish this limit at −273.16°C, or −459.69°F. As this represents a lower limit, temperature scales are readjusted to assign zero to this limit. The *absolute scales* in use are the Kelvin scales (the size of the degree is equivalent to the centigrade degree), in which all temperatures are 273.16 degrees higher. In the Rankine scale, the size of the degree is the same as that of the Fahrenheit degree, but all temperatures are 459.69 degrees higher.

Numerous other quantities enable the properties of matter to be character-

Table 1.4. Commonly Used Symbols and Definitions.

General Symbols

	Symbol	Unit or Definition
Acceleration	a	m/s²
of Gravity	g	m/s²
Base of Natural Logarithms	e	
Coefficient	C	
Difference, finite	Δ	
Differential Operator	d	
Partial	δ	
Efficiency	η	
Energy, dimension of	E	J, N-m
Enthalpy	H	J
Entropy	S	J/K
Force	F	N
Function	Φ, ψ, χ	
Gas Constant, universal	R	To distinguish, use R_o
Gibbs Free Energy	G, F	G = H − TS, J
Heat	Q	J
Helmholtz Free Energy	A	A = U − TS, J
Internal Energy	U	J
Mass, dimension of	m	kg
Mechanical Equivalent of Heat	J	Unity, dimensionless
Moment of Inertia	I	(m)⁴
Newton's Law of Motion, conversion factor in	g_c	Unity, dimensionless
Number		
In general	N	
Of moles	n	
Pressure	p, P	Pa, bar
Quantity, in general	Q	
Ratio, in general	R	
Resistance	R	
Shear Stress	τ	Pa
Temperature		
Dimension of	θ	
Absolute	T	K (Kelvin)
In general	T, t	°C
Temperature Difference, logarithmic mean	$\bar{\theta}$	°C
Time		
Dimension of	T	s
In general	t, τ	s, hr
Work	W	J

Geometric Symbols

Linear dimension		
Breadth	b	m
Diameter	D	m
Distance along Path	s, x	m
Height above Datum Plane	Z	m
Height Equivalent	H	m
Hydraulic Radius	r_H, R_H	m, m²/m

Table 1.4. (continued).

Geometric Symbols *(continued)*		
	Symbol	**Unit or Definition**
Lateral Distance from Datum Plane	Y	m
Length, distance or dimension of	L	m
Longitudinal Distance from Datum Place	X	m
Mean Free Path	λ	m
Radius	r, R	m
Thickness		
In general	B	m
Of file	B_f	m
Wavelength	λ	m
Area		
In general	A	m²
Cross section	S	m²
Fraction-free Cross Section	σ	
Projected	A_p	m²
Surface		
per unit mass	A_w, s	m²/kg
Volume		
In general	V	m³
Fraction voids	ϵ	
Humid volume	ν_H	m³/kg dry air
Angle	α, θ, Φ	
In x, y plane	α	
In y, z plane	ϕ	
In z, x plane	θ	
Solid angle	ω	
Other		
Particle-shape factor	ϕ_s	

Intensive Properties Symbols		
Absorptivity for Radiation	α	
Activity	a	
Activity Coefficient, molal basis	γ	
Coefficient of Expansion		
Linear	α	m/(m-K)
Volumetric	β	m³/(m³-K)
Compressibility Factor	z	$z = pV/RT$
Density	ϱ	kg/m³
Diffusivity		
Molecular, volumetric	D_v, δ	m³/(s-m), m²/s
Thermal	α	$\alpha = k/C_p$, m²/s
Emissivity Ratio for Radiation	e	
Enthalpy, per mole	H	J/kmol
Entropy, per mole	S	J/(kmol-K)
Fugacity	f	Pa, bar
Gibbs Free Energy, per mole	G, F	J/kmol
Helmholtz Free Energy, per mole	A	J/kmol
Humid Heat	c_s	J/(kg dry air-K)

(continued)

Table 1.4. (continued).

Intensive Properties Symbols *(continued)*

	Symbol	Unit or Definition
Internal Energy, per mole	U	J/kmol
Latent Heat, phase change	λ	J/kg
Molecular Weight	M	kg
Reflectivity for Radiation	ϱ	
Specific Heat	c	J/(kg-K)
At constant pressure	c_p	J/(kg-K)
At constant volume	c_v	J/(kg-K)
Specific Heats, ratio of	γ	
Surface Tension	σ	N/m
Thermal Conductivity	k	(J-m)/(s-m²-K)
Transmissivity of Radiation	τ	
Vapor Pressure	p*	Pa, bar
Viscosity		
Absolute or coefficient of	μ	Pa-s
Kinematic	ν	m²/s
Volume, per mole	V	m³/kmol

Symbols for Concentrations

	Symbol	Unit or Definition
Absorption Factor	A	$A = L/K^*V$
Concentration (mass or moles per unit volume)	c	kg/m³, kmol/m³
Fraction		
Cumulative beyond a given size	ϕ	
By volume	χ_v	
By weight	χ_μ	
Humidity	H, Y_H	kg/kg dry air
At saturation	H_s, Y^*	kg/kg dry air
At wet-bulb temperature	H_w, Y_w	kg/kg dry air
At adiabatic saturation	H_a, Y_a	kg/kg dry air
Mass Concentration of Particles	c_p	kg/m³
Moisture Content		
Total water to bone-dry stock	$X\phi^*$	kg/kg dry stock
Equilibrium water to bone-dry stock	X^*	kg/kg dry stock
Free water to bone-dry stock	X	kg/kg dry stock
Mole or Mass Fraction		
In heavy or extract phase	x	
In light or raffinate phase	y	
Mole or Mass Ratio		
In heavy or extract phase	X	
In light or raffinate phase	Y	
Number Concentration of Particles	n_p	number/m³
Phase Equilibrium Ratio	K^*	$K^* = y^*/x$
Relative Distribution of Two Components		
Between two phases in equilibrium	α	$\alpha = K_i^*/K_j^*$
Between successive stages	β	$\beta = (y_i/\nu_i)_x/(x_i x_i)_{n+1}$
Relative Humidity	H_R, R_H	
Slope of Equilibrium Curve	m	$m = dy^*/dx$
Stripping Factor	S	$S = K^*V/L$

Table 1.4. (continued).

Rate Symbols

	Symbol	Unit or Definition
Quantity per Unit Time, in general	q	
Angular velocity	ω	
Feed rate	F	kg/s, kmol/s
Frequency	f, N_f	
Friction Velocity	u*	$u^* = (\tau_w \varrho)^{1/2}$, m/s
Heat Transfer Rate	q	J/s
Heavy or Extract Phase Rate	L	kg/s, kmol/s
Heavy or Extract Product Rate	B	kg/s, kmol/s
Light or Raffinate Phase Rate	V	kg/s, kmol/s
Light or Raffinate Product Rate	D	kg/s, kmol/s
Mass Rate of Flow	w	kg/s, kg/hr
Molal Rate of Transfer	N	kmol/s
Power	P	W
Velocity, in general	n	m/s
Revolutions per Unit Time	u	m/s
Longitudinal (x), component of	u	m/s
Lateral (y), component of	v	m/s
Normal (z), component of	w	m/s
Volumetric Rate of Flow	q	m³/s, m³/hr
Quantity per Unit Time, Unit Area		
Emissive power, total	W	W/m²
Mass velocity, average	G	G = w/S, kg/(s-m²)
Vapor or light phase	G, $\bar{G}$	kg/(s-m²)
Liquid or heavy phase	L, $\bar{L}$	kg/(s-m²)
Radiation, intensity of	I	W/m²
Velocity		
Nominal, basis total cross section of packed vessel	v_s	m/s
Volumetric average	V, $\bar{V}$	m³/(s-m²), m/s
Quantity per Unit Time, Unit Volume		
Quantity reacted per unit time, reactor volume	N_R	kmol/(s-m²)
Space Velocity, volumetric	Λ	m³/(s-m³)
Quantity per Unit Time, Unit Area, Unit Driving Force, in general	k	
Eddy Diffusivity	δ_E	m²/s
Eddy Viscosity	ν_E	m²/s
Eddy Thermal Diffusivity	α_E	m²/s
Heat Transfer Coefficient		
Individual	h	W/(m²·K)
Overall	U	W/(m²·K)
Mass Transfer Coefficient		
Individual	k	kmol/(s-m²) (driving force)
Gas film	k_G	To define driving force use subscript:
Liquid film	k_L	c for kmol/m³
Overall	K	p for bar
Gas film basis	K_G	x for mole fraction
Liquid film basis	K_L	
Stefan-Boltzmann Constant	σ	5.6703 × 10⁻⁸ W/(m²·K⁴)

ized. Most are elementary and require no discussion. However, some of the thermodynamic quantities such as internal energy, work, and enthalpy should be reviewed briefly before tackling engineering problems.

First, we note the quantities of volume and pressure. Volume, like mass, depends on the amount of material under examination. In contrast, specific volume is defined as the volume per unit mass or volume per mole of material which is thus independent of the amount of material.

The pressure of a fluid is defined as the normal force exerted per unit area of surface ($p = F_0/F$, where F is area and F_0 force). Pressure is measured most frequently in terms of the height of a column of fluid under the influence of gravity. For example, a mercury manometer is used to obtain pressure measurements in millimeters of mercury. Such values are converted to force per unit area by multiplying the height of the column by the fluid's density. That is, the force exerted by gravity on the mercury column is

$$F_0 = \frac{1}{g_c}(m)(g) \quad (1.3)$$

where g is the local acceleration of gravity and $m = (F)(h)(g)$. By eliminating mass, m, we obtain

$$\frac{F_0}{F} = p = (h)(\varrho)\frac{g}{g_c} \quad (1.4)$$

Since the density of the manometric fluid is a function of temperature, the pressure corresponding to a given height of fluid column also depends on temperature. A common unit of pressure is the atmosphere, which corresponds to the force of gravity acting on air above the earth's surface. A standard atmosphere is equal to 14.696 lb_f/in^2, or 29.921 in Hg at 0°C in a standard gravitational field. Note that in SI a *bar* is equivalent to 0.9869 atm or 10^6 dyne/cm². In many cases, measurements obtained represent a difference between a system under study and the pressure of the surrounding environment. These measurements are known as *gauge pressures,* which can be converted to *absolute* pressures by adding the barometric pressure.

Work is performed when a force acts through a distance. It is defined as the product of the force and the distance over which it is applied. If the force is constant, then

$$W = F_0 S \quad (1.5A)$$

If it is variable, then

$$dW = F_0 dS \quad (1.5B)$$

To determine the work for a finite process, Equation (1.5B) must be integrated over appropriate limits. As an example, consider the compression or expansion of a gas by the movement of a piston. The distance over which the piston moves is equal to the volume change of the gas divided by the area of the piston, or

$$dW = (pF)d\,\frac{V}{F}$$

with constant F

$$dW = pdV$$

or

$$W = \int_{V_1}^{V_2} pdV \qquad (1.6)$$

This is the general expression for work performed as a result of a finite compression or expansion process.

From Equation (1.5B), a body moves through a differential distance dS over some differential time $d\theta$. Denoting u as the body velocity, then

$$a = \frac{du}{d\theta}$$

whence

$$dW = \frac{m}{g_c}\,udu$$

Integrating over a finite change in velocity,

$$W = \frac{m}{g_c}\int_{u_1}^{u_2} udu$$
$$= \frac{mu_2^2}{2g_c} - \frac{mu_1^2}{2g_c} = \Delta\frac{mu^2}{2g_c} \qquad (1.7)$$

The term $mu^2/2g_c$ is termed the *kinetic energy* of the body. The work performed on a body by accelerating it from an initial velocity u_1 to some final velocity u_2 is equivalent to the change in the body's kinetic energy.

When a body of mass is elevated from an initial position z_1 to a final level z_2, an upward force that is equivalent to at least the weight of the body must be exerted on it. This force must move through the distance z_2-z_1; as the weight of the body is the force of gravity on it, the minimum force required is given by Equation (1.3). That is, the minimum amount of work required to raise the body is

$$W = F_0(z_2 - z_1)$$
(1.8)
$$= mz_2 \frac{g}{g_c} - mz_1 \frac{g}{g_c} = \Delta \frac{mzg}{g_c}$$

This is the work done in elevating a body or its *potential energy*. If the elevated body is permitted to fall freely, it will gain in kinetic energy what it loses in potential energy. Hence, the capacity for doing work remains unchanged:

$$\Delta KE = \Delta PE = 0$$
(1.9)

This is known as the principle of conservation of energy in mechanics and is a true statement provided that both kinetic and potential energy changes are equivalent to the work done in producing them. Work is essentially energy in transit. When work is performed and does not simultaneously appear elsewhere in the system, it is converted into another form of energy. In discussing specific unit operations, the body on or by which work is performed is referred to as the *system*. When work is performed, it is done by the surroundings on the system, or vice versa. Thus, energy is transferred from the surroundings on the system, or vice versa. Thus, energy is transferred from the surroundings to the system, or the reverse. The form of energy referred to as work exists during this transfer. On the other hand, kinetic and potential energies reside with the system. They are measured with reference to the surroundings, and changes are independent of this reference or datum level.

Internal Energy and the First Law of Thermodynamics

The internal energy of a substance refers to the energy of the molecules composing the material. Molecules are in perpetual motion and thus possess kinetic energies of translation, rotation, and vibration. The addition of heat increases the molecular activity and, consequently, causes an increase in the internal energy. Work performed on a substance has the same effect. In addition to kinetic energy, molecules also possess potential energy, which results from forces of attraction between them. Hence, internal energy is from within the system, whereas kinetic and potential energies are external forms.

As heat and internal energy are both forms of energy, the general law of conservation of energy (i.e., the first law of thermodynamics) may be stated in this way: *regardless of the form of energy, the total quantity of energy is constant, and when energy in one form vanishes, it appears simultaneously in other forms.* In relation to the system and surroundings, this is

$$\Delta(\text{energy of the system}) + \Delta(\text{energy of the surroundings}) = 0 \quad (1.10)$$

The first term can be expanded to show energy changes in different forms:

$$\Delta(\text{energy of the system}) = \Delta U + \Delta KE + \Delta PE \quad (1.11)$$

or

$$\Delta U + \Delta KE + \Delta PE = \pm Q \pm W \quad (1.12)$$

where ΔU, ΔKE and ΔPE represent changes in internal, kinetic, and potential energies of the system, respectively, and Q and W are heat and work, respectively. By convention, heat is regarded as positive when it is transferred to the system from the surroundings, whereas work is taken to be positive when transferred from the system to the surroundings.

In many cases, the system does not undergo a change in kinetic and potential energy, but only in internal energy. In these special cases,

$$dU = dQ - dW \quad (1.13)$$

where the expression has been written to show differential changes.

From the standpoint of thermodynamics, there are two types of quantities: (1) those that depend on the path taken to achieve a final state, and (2) those that do not depend on the past history of the substance or on the path followed in reaching a given state. In the latter case, properties depend on the immediate conditions and are referred to as point functions or *state functions*.

Values of state functions may be represented as individual points on a graph. The differential of a state function represents an infinitesimal change in property. The integration of such differentials yields a finite difference between two values of the specific property, e.g.,

$$\int_{u_1}^{u_2} dU = U_2 - U_1 = \Delta U$$

In contrast, work and heat are not state properties but are functions of the path followed. They are not represented by points on a graph but rather are denoted by areas. It should be noted also that a state function represents an instantaneous property of a particular system. In contrast, work and heat exist only when changes are caused in a system. This implies further that whenever heat is transferred or work performed, an increment of time occurs. The quantities of heat and work are critical in the design of unit operations equipment such as heat exchangers, distillation columns, evaporators, pumps, compressors, turbines, etc. State functions, such as internal energy, are properties of matter; however, once measured, they may be applied to calculating Q and W for any process through the laws of thermodynamics.

Enthalpy

Enthalpy is a thermodynamic function defined by the following relation:

$$dH = dU + d(pV) \qquad (1.14)$$

where U is the internal energy of the system, p the absolute pressure, and V the system volume. As U, p, and V are all state functions, so is H. Equation (1.14) applies whenever a differential change to the system occurs. The equation may be expressed for any amount of material; however, it is usually expressed on a unit mass or mole basis. The importance of this thermodynamic concept to engineering problems will become apparent in the chapters to follow.

Heat Capacity and Specific Heat

Two additional thermodynamic terms of importance are a material's heat capacity and specific heat. The flow of heat is commonly thought of in terms of its effects on the substances that receive the energy. Heat capacity refers to the quantity of heat required to raise the temperature of a given mass by 1°. By convention, this quantity is based on either 1 mole or a unit mass of material. The mathematical relationship is

$$dQ = nCdT \qquad (1.15A)$$

where n is the number of moles, C is the molal heat capacity, and dT is the temperature increment resulting from the quantity of heat dQ. On a unit mass basis, this expression becomes

$$dQ = mCdT \qquad (1.15B)$$

where m is mass, and C is per unit mass.

By rigorous definition, the term specific heat is the ratio of the heat capacity of a material to the heat capacity of an equal quantity of water. The specific heat of water is approximately 1 cal/(g)(°C); the molal heat capacity of water is 18 cal/(g-mol)(°C).

Let us assume a quantity of gas is contained in a vessel with rigid walls. On the addition of heat, the gas temperature rises, but the system volume remains constant. Then

$$dQ = C_v dT$$

or

$$Q = C_v \Delta T$$

As the system did not undergo a change in volume, no work was performed, and it is easy to show that the first law of thermodynamics reduces to

$$\Delta U = Q = C_v \Delta T \tag{1.16}$$

In contrast, assume the same operation is now performed as follows: a quantity of gas is retained in a cylinder by a frictionless piston at constant pressure p. If heat is added, the gas will expand reversibly, and if the force on the piston is maintained constant, the process takes place at constant p. Then

$$Q = C_p \Delta T$$

As the gas expands, work is performed, and the first law expression reduces to

$$\Delta H = C_p \Delta T = Q \tag{1.17}$$

The student should prove the definition. Thus, for a constant-pressure process carried out reversibly, the system's enthalpy change is equal to the heat added. For purposes of calculating changes in properties, the equation applying to the reversible constant-pressure process may be applied to actual or reversible processes that accomplish the same change in state [see Smith and Van Ness (1959)].

Properties of the Ideal Gas

Gases seldom, if ever, are handled at standard conditions (i.e., STP, standard temperature and pressure = 0°C, 1 atm). Consequently, it is necessary to estimate the molal volume at different temperatures and pressures. For a so-called ideal gas, a simple p-V-T relationship exists. An ideal gas is one characterized by the following:

1 The volume of individual molecules is negligible in comparison to the total volume of the gas.
2 No forces exist between the molecules of the gas.

Applying these assumptions to the kinetic theory of gases [see Cheremisinoff (1981) or Bird et al. (1960) for derivation], the *ideal gas law* is obtained:

$$pV = nRT \qquad (1.18A)$$

where

V = volume of gas
p = absolute pressure
T = absolute temperature
R = universal gas law constant
n = number of moles of gas

$$p\dot{v} = RT \qquad (1.18B)$$

where $\dot{v}$ is the gas volume per mole.

For most gases at pressures of a few atmospheres, the ideal gas law may be applied to engineering calculations. Various values of the constant R in different units are as follows:

R = 1.987 Btu/(lb-mol)(°F) or cal/(g-mol)(°K)
   = 0.730 (atm)(ft^3)/(lb-mol)(°R)
   = 10.73 (psi)(ft^3)/(lb-mol)(°R)
   = 1.314 (atm)(ft^3)/(lb-mol)(°R)
   = 1545 (lb$_f$/ft^2)(ft^3)/(lb-mol)(°R) or ft-lb$_f$/(lb-mol)(°R)
   = 82.06 (atm)(cm^3)/(g-mol)(°R)

The characteristics of ideal gases are best described in terms of the thermodynamics of commonly encountered situations. Therefore, we examine the behavior of 1 mole of ideal gas in a reversible nonflow process for five cases: isometric (constant-volume), isobaric (constant-pressure), isothermal (constant-temperature), adiabatic, and polytropic.

In a *constant-volume (isometric) process,* Equation (1.16) is directly applicable to an ideal gas. In integral form, this expression is

$$\Delta U = Q = \int C_v dT$$

The equation states that the internal energy of an ideal gas is only a function of temperature (based on characteristic 2 above). As U is independent of the specific volume at constant temperature, a plot of U versus $\dot{v}$ will result in a straight line. For different temperatures, U has different values.

For a *constant-pressure (isobaric) process,* the first law expression developed earlier applies:

$$dH = dQ = C_p dT$$

or, for constant C_p,

$$\Delta H = Q = C_p \Delta T$$

As the energy of an ideal gas depends only on temperature, its enthalpy is also strictly a function of T. This is obvious from

$$dH = dU + d(p\dot{v})$$

or

$$dH = dU + RdT \text{ (for an ideal gas)} \qquad (1.19)$$

Equation (1.19) provides a relationship between C_p and C_v:

$$C_p dT = C_v dT + RdT$$

whence

$$C_p = C_v + R \qquad (1.20)$$

$$\left. \begin{array}{l} dU = dQ - dW = 0 \\ \\ Q = W \end{array} \right\} \qquad (1.21)$$

or

Thus, for an ideal gas, we may write

$$Q = W = \int p \, d\dot{v} = \int RT \frac{d\dot{v}}{\dot{v}} \qquad (1.22)$$

On integration at constant temperature, we obtain

$$\left. \begin{array}{l} Q = W = RT \ln \dfrac{\dot{v}_2}{\dot{v}_1} \\ \\ = RT \ln \dfrac{P_1}{P_2} \end{array} \right\} \qquad (1.23)$$

because $\dot{v}_2/\dot{v}_1 = P_1/P_2$.

An *adiabatic process* is one in which there is no transfer of heat between system and surroundings, i.e., $dQ = 0$. For a reversible adiabatic process then,

$$dU = -dW = -p \, d\dot{v} \qquad (1.24)$$

And because $C_v dT = -pd\dot{v}$ and $p = RT/\dot{v}$, we obtain

$$\frac{dT}{T} = -\frac{R}{C_v}\frac{d\dot{v}}{\dot{v}} \tag{1.25}$$

Denoting γ as the ratio of heat capacities, i.e., $\gamma = C_p/C_{\dot{v}} = (C_{\dot{v}} + R)/C_{\dot{v}} = 1 + R/C_{\dot{v}}$, Equation (1.25) may be rewritten as follows:

$$\frac{dT}{T} = -(\gamma - 1)\frac{d\dot{v}}{\dot{v}} \tag{1.26}$$

Integrating this expression gives

$$\left(\frac{T_2}{T_1}\right) = \left(\frac{\dot{v}_1}{\dot{v}_2}\right)^{\gamma-1} \tag{1.27A}$$

And through the ideal gas law relation, we also obtain

$$\left(\frac{T_2}{T_1}\right) = \left(\frac{P_2}{P_1}\right)^{(\gamma-1)/\gamma} \tag{1.27B}$$

$$\left(\frac{\dot{v}_1}{\dot{v}}\right)^{\gamma-1} = \left(\frac{P_2}{P_1}\right)^{(\gamma-1)/\gamma} \tag{1.27C}$$

or, finally,

$$P_1\dot{v}_1^\gamma = P_2\dot{v}_2^\gamma = P\dot{v}^\gamma = \text{constant} \tag{1.28}$$

Leaving the details of the derivation to the student, we write the expression for work of an adiabatic process $(-dW = dU = C_{\dot{v}}dT)$ as

$$W = \frac{P_1\dot{v}_1 - P_2\dot{v}_2}{\gamma - 1} \tag{1.29}$$

Or, rewriting this expression in terms of $\dot{v}_1$ only,

$$W = \frac{P_1\dot{v}_1}{\gamma - 1}\left[1 - \left(\frac{P_2}{P_1}\right)^{(\gamma-1)/\gamma}\right] = \frac{RT_1}{\gamma - 1}\left[1 - \left(\frac{P_2}{P_1}\right)^{(\gamma-1)/\gamma}\right] \tag{1.30}$$

Finally, a *polytropic process* is a general case in which no specific conditions other than reversibility are imposed. The above general equations for nonflow processes apply in this case to ideal gases:

$$dU = dQ - dW, \qquad \Delta U = Q - W$$

$$dW = pd\dot{v}, \qquad W = \int pd\dot{v}$$

$$dU = C_v dT, \qquad \Delta U = \int C_v dT$$

$$dH = C_p dT, \qquad \Delta H = \int C_p dT$$

and

$$dQ = C_v dT + pd\dot{v}$$
$$Q = \int C_v dT + \int pd\dot{v} \tag{1.31}$$

Work may be calculated directly from the integral $\int pd\dot{v}$.

Note that the expressions presented in this section were derived for reversible nonflow processes involving ideal gases. Those expressions, however, which only relate state functions, are applicable to ideal gases for both reversible and irreversible flow and nonflow processes because changes in state functions depend only on the initial and final states of the system. However, expressions for Q and W are specific to the cases considered.

BEHAVIOR OF FLUIDS

Unlike gases, which are nature's simplest substances, liquids are more complex in behavior and, hence, cannot be described by a simple thermodynamic relation. The complexity of liquid properties is due to their high density in comparison to the low molecular density of gases. The complex thermodynamic nature of liquids is beyond the scope of this text. For a discussion of the equations of states for liquids and transport properties, the reader is referred to the work of Sridhar (1983). Rather, we shall describe briefly the physical behavior of fluids (liquids) when subjected to a force.

A *fluid* (gas or liquid) is any form of matter that deforms continuously under the action of a shearing stress. When this stress is removed, deformation ceases, but the fluid does not return to its original configuration. This behavior is in contrast to that of a *solid,* which does not deform continuously under a shearing stress, but attains a definite equilibrium deformed state for a particular stress. On removal of the stress, the solid returns to its original configuration (assuming that the solid is being stressed below its yield value).

In practice, such an absolute definition is not applied to a fluid. There are many materials that exhibit both fluid-like and solid-like properties but are still considered to be fluids. For example, certain materials will not deform continuously under the action of any nonzero shearing stress until a certain "yield" stress is exceeded. Still other materials deform continuously, but when the stress is removed they partially recover their original configuration. As a compromise, we may consider a fluid to be a form of matter that exhibits continuous deformation under some range of shearing stress and may partially recover its original configuration when the stress is removed.

The deformation of a fluid under the action of a stress is the subject of the science of *rheology*. The study of the deformation and flow of materials is a complex subject that is still being developed. Extensive reviews of the subject have been done by Bird et al. (1977), Fredrickson (1964), Lodge (1964), Middleman (1968), and Schowalter (1978), among others.

Rheologists are concerned with the measurement of the deformation of a fluid with stress and the formulation of mathematical relations between deformation rates (or velocity gradients) and stress τ. These relations are referred to as rheological models, or *constitutive equations*, which can be used in a momentum balance equation to compute for a given laminar flow situation, velocity profiles, volumetric flow rates, etc. While a large number of such equations have been formulated, their use often entails the evaluation of multiple parameters from data, which is difficult in practice. Furthermore, the complexity of many of these equations precludes their use in engineering design. Unfortunately, parameters evaluated from measurements under steady shear often cannot represent data from, say, oscillatory shear. As such, only very simple constitutive equations are applied in engineering work. Obviously, these simple rheological models are limited in their ability to represent all aspects of fluid behavior and are suitable only under restricted conditions.

Consider a fluid contained between two large parallel plates separated by a small gap Y, as shown in Figure 1.1. The lower plate is stationary, while the upper plate is moved at a constant velocity v through the action of applied force F_0. A thin layer of fluid adjacent to each plate will move at the same velocity as the plate (referred to as the "no slip" condition which holds true for all except a few fluids). Molecules in the fluid layers between these two extremes will move at intermediate velocities. For example, a fluid layer B immediately below A will experience a force in the x direction from layer A and a smaller retarding force from layer C. Layer B then will flow at a velocity lower than v. This progression continues to the layer adjacent to the lower plate, whose velocity is zero. Under steady-state conditions, the force F_0 required to produce motion becomes constant and is related to velocity as follows:

$$\frac{F_0}{F} = \psi\left(\frac{v}{Y}\right) \qquad (1.32)$$

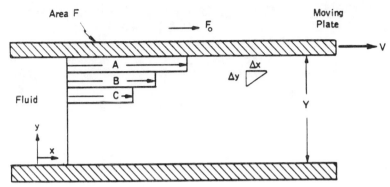

FIGURE 1.1. The deformation of a fluid.

where F is the area of each plate, and ψ is a function of the fluid properties, temperature, and pressure.

The ratio F_o/F is the shear stress on the fluid at the upper plate, and v/Y is the velocity gradient. On a local basis, that is, at any point within a fluid, the above expression may be generalized:

$$\tau_{yx} = \psi\left(\frac{dv_x}{dy}\right) \qquad (1.33)$$

τ_{yx} is the local shear stress (or force per unit area) acting in the x direction on a plane perpendicular to the y axis; v_x is the velocity in the x direction; dv_x/dy is the local velocity gradient. Equation (1.33) represents the constitutive equation for the fluid in question. In graphic form it is referred to as the flow curve, or *rheogram* for the fluid.

It is instructive to interpret the velocity gradient as follows:

$$\frac{dv_x}{dy} = \frac{d}{dy}\left(\frac{dx}{dt}\right) = \frac{d}{dt}\left(\frac{dx}{dy}\right) \approx \frac{d}{dt}\left(\frac{\Delta x}{\Delta y}\right) \qquad (1.34)$$

The term $\Delta x/\Delta y$ is the shear strain on the fluid, whereas dv_x/dy is the rate of shear strain or, simply, the shear rate. Hence, for fluids the shear stress is a function of the rate of shear strain or, simply, the shear rate. In contrast, for solids the shear stress is a function of strain rather than the rate of strain. Shear rate is often called the rate of deformation and is denoted by the symbol $\dot{\gamma}$.

For a homogeneous fluid containing small molecules, the constitutive Equation (1.33) is usually simple, but for a multiphase mixture—a solution or a liquid containing large molecules—complex relations result. In the first case, the shear stress–shear rate relation is linear through the origin. Such fluids are termed Newtonian. In these simple fluids, the internal structure is unaffected by the magnitude of the imposed shear rate. In complex fluids, an imposed shear results in changes in the internal structure of the fluid so that the constitutive relation becomes more complicated. For example, in a liquid containing large molecules (a molten polymer or a polymer solution), at low shear rates the molecules remain randomly coiled, much as they are in the fluid at rest. The fluid structure remains unchanged in this range, and the shear stress–shear rate relation is linear as for a Newtonian fluid. However, as the shear rate is increased, the randomly coiled molecules tend to line up in the flow direction, changing the structure of the fluid and, hence, the nature of the constitutive relationship. The progressive lining-up or disentangling of the molecules results in the fluid becoming less viscous, or "thinner," in its flow properties. Increasing the shear further would cause more and more molecules to line up, so that the stress–shear rate relation would remain nonlinear until a limiting shear is reached when all the molecules have lined up. Further shear rate increases would not result in structural change, and the constitutive relation would return to a Newtonian relation (albeit a different one from that for low shear rates). Thus, the progression is from Newtonian to non-Newtonian and back as the shear rate is increased. Structural changes also occur for suspensions of solid particles in a liquid or for liquid–liquid emulsions, resulting again in complex flow curves for such fluids.

Under the condition of steady flow as developed by the system in Figure 1.1, time-dependent properties and solid-like or elastic behavior cannot be detected. However, consider what would occur if the force moving the upper plate were removed. In most instances, the upper plate would continue to move, but with decreasing velocity until it and the fluid eventually come to rest. If the fluid were *viscoelastic,* i.e., possessing both viscous and elastic properties, the plate and the fluid would first slow down as before; however, after coming to a stop, some motion in the negative x direction would occur as the fluid sought to recover its original configuration. Only partial recovery would be attained. Next, consider what happens if the shear rate on a fluid is changed instantaneously—a situation difficult to achieve experimentally because in our system the inertia of the plate and the fluid itself will result only in a gradual change. For many fluids, any resulting changes in internal structure occur very rapidly, and a new stress level corresponding to the new shear rate is reached instantaneously following Equation (1.33). Such fluids are termed *time-dependent*. For *time-dependent* fluids, structural changes are considerably slower, and the shear stress changes slowly until ultimately a steady value is reached corresponding to the new shear rate. Some fluids do not deform until a certain yield stress vlaue is exceeded. In the experiment in Figure 1.1, F_0/F would have to exceed the yield stress for flow to occur for such a liquid.

Fluids may be divided into three classes based on their flow behavior: time-independent purely viscous fluids, time-dependent purely viscous fluids, and viscoelastic fluids. These classes are summarized in Table 1.5. Of these three classes, the time-independent purely viscous fluids are understood best. A great deal of work has been done on viscoelastic fluids, but few results of use in engineering design are available. Time-dependent fluids are less common and perhaps the least understood.

Time-Independent Fluids

Newtonian Fluids

As noted above, the simplest class of real fluids comprises the Newtonians, whose constitutive equation is given by

$$\tau = \mu \dot{\gamma} \qquad (1.35A)$$

where, for convenience, we have dropped the yx subscripts on τ. Equation (1.35A) is known as Newton's law of viscosity. Viscosity μ is a property of the fluid that is a function of temperature and pressure only (for a single-phase, single-component system). With temperature, the viscosity of liquids decreases, whereas that of gases increases. An *ideal* or perfect fluid is one whose viscosity is zero and one that can be sheared without the applicaton of a shear stress. An ideal fluid does not exist, but the concept is useful in the theory of potential flows.

The dimensions of μ are readily developed. The stress τ has dimensions ML^{-1} and $\dot{\gamma}$, t^{-1}. Thus, μ has dimensions $ML^{-1}t^{-1}$. In the cgs units system (M in g, L in cm, t in s), the viscosity has units g cm^{-1} s^{-1}, known as poise. For most Newtonian fluids this is a rather large unit, and the centipoise (cp), which is 0.01 poise, is more convenient. For points of reference, the viscosity of water at 20°C is 1.00 cp and that of air is 0.0181 cp.

Table 1.5. Classification of Fluid Behavior.

Fluids				
Purely Viscous				
Time Independent		Time Dependent		
No Yield Stress	Yield Stress	Thixotropic		Viscoelastic
Newtonian Pseudoplastic Dilatant	Bingham Yield-Pseudoplastic Yield-Dilatant	Rheopectic		

As noted, the flow curve or rheogram of a Newtonian fluid is a straight line through the origin in arithmetic coordinates. The slope of the line is the viscosity, so that the entire class of Newtonian fluids can be represented by a family of straight lines through the origin. It is often convenient to use logarithmic coordinates for plotting the rheogram, because they allow a greater range of data and also permit easier comparison of Newtonian versus non-Newtonian behavior. On such coordinates, a Newtonian fluid has a rheogram that is a straight line of unit slope and whose intercept at a shear rate of unity is the viscosity. The plots in Figures 1.2(a) and 1.2(b) illustrate these alternative methods of representation.

Most low-molecular-weight liquids and solutions, and all gases are Newtonian. Homogeneous slurries of small spherical particles in gases or liquids at low solids concentration are also frequently Newtonian. Thus, there is a large class of fluids for which the property of viscosity characterizes flow behavior.

A far larger class of fluids *does not* follow Newton's law of viscosity and, furthermore, may display other characteristics such as time-dependence, viscoelasticity, and yield. A fluid with any of these characteristics is termed non-Newtonian. In general, there are no completely satisfactory constitutive equations for non-Newtonian fluids, in contrast to the situation for Newtonian fluids.

For non-Newtonian fluids, it is convenient to define an "apparent viscosity" η_a as

$$\eta_a = \tau/\dot{\gamma} \qquad (1.35B)$$

The apparent viscosity is a function of $\dot{\gamma}$ for non-Newtonian materials and is analogous to the Newtonian viscosity μ. However, whereas the Newtonian viscosity does not vary with shear rate, the non-Newtonian apparent viscosity does.

Pseudoplastic or Shear-Thinning Fluids

Pseudoplastic fluids are often referred to as shear-thinning, because their apparent viscosity decreases with shear rate. That is, the rate of increase of shear stress for such fluids decreases with increased shear rate. Increased shear rapidly breaks down the internal structure within the fluid and does so reversibly, whereby no time-dependence is manifested. Examples of fluids that exhibit shear-thinning are polymer melts and solutions, mayonnaise, suspensions such as paint and paper pulp, and some dilute suspensions of inert particles. Many of these fluids also exhibit other non-Newtonian characteristics such as viscoelasticity, in the case of polymer solutions and melts, and time-depencence, in the case of paints. Thus, pseudoplasticity is but one important characteristic of such a non-Newtonian fluid and does not

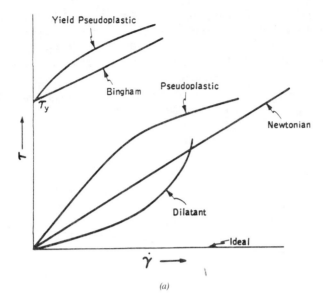

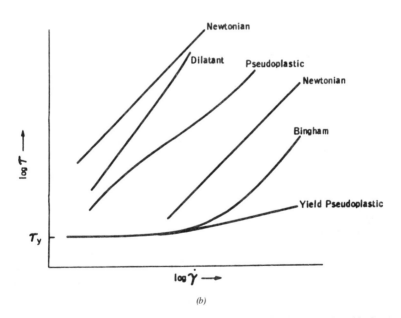

FIGURE 1.2. (a) Rheograms on arithmetic coordinates; (b) rheograms on logarithmic coordinates.

necessarily describe all its non-Newtonian features. Many pseudoplastics are shear-thinning at intermediate shear rates and are Newtonian at low and high shear rates. A mechanism for this phenomenon for fluids containing long molecules such as polymer solutions was noted above. Rheograms for pseudoplastics are shown in Figure 1.2.

Many constitutive equations of varying complexity have been proposed for these fluids. Because of its simplicity, a power law relation is used most widely, even though it does not describe the Newtonian extremes found for most pseudoplastics. The power law, or Ostwald-de Waele equation:

$$\tau = K\dot{\gamma}^n \tag{1.36}$$

is a two-parameter equation. Parameter K is called the *consistency index* and n is the *flow behavior index*. Both are functions of temperature and pressure, but K is more sensitive to temperature than n. Pressure dependence of these parameters has not been investigated.

Note that if $n = 1$, Equation (1.36) reduces to Newton's law of viscosity. For $n < 1$, pseudoplastic or shear-thinning behavior is exhibited, whereas if $n > 1$, dilatant or shear-thickening behavior is observed. The apparent viscosity for a power law fluid is

$$\eta_a \equiv \tau/\dot{\gamma} = K\dot{\gamma}^{n-1} \tag{1.37}$$

showing that if $n < 1$, η_a decreases with $\dot{\gamma}$. The value of n is the slope of the rheogram on logarithmic coordinates.

A fit of the power law expression to experimental dat is usually good over several orders of magnitude in $\dot{\gamma}$. However, at low and high $\dot{\gamma}$, Newtonian behavior often persists and agreement with the power law is poor.

Note that Equation (1.36) is valid only for $\dot{\gamma} > 0$. Depending on the coordinate system, $\dot{\gamma}$ may be negative over some portion of the flow field, in which case the correct form of this equation is

$$\tau = K|\dot{\gamma}|^{n-1}\dot{\gamma} \tag{1.38}$$

In this book we assume $\dot{\gamma} > 0$, and "simple shear" contains only one nonzero component of the velocity gradient. Extensions of Equation (1.37) to more complicated flow fields and where $\dot{\gamma}$ may be negative are given by Bird et al. (1960). These authors place a negative sign in the constitutive equations because of their interpretation of τ as momentum flux rather than shear stress. This sign should be ignored to conform to the notation in this text.

Other equations have been proposed for pseudoplastics but have not been used as widely as the power law. Also, they are generally more complex, and some involve three or more parameters, rather than only two. The principal advantage of these more complex models is that they can predict shear-thinning behavior as well as the tendency to Newtonian behavior at one or both extremes of shear rate. Table 1.6 lists some of these models along with their main features. Further details are given by Fredrickson (1964), Skelland (1967), and Bird et al. (1960).

Dilatant or Shear-Thickening Fluids

Dilatant fluids (also called shear-thickening fluids) are those whose apparent viscosity increases with shear rate. The rate of increase of shear stress for such fluids increases with shear rate. From appropriate values of the parameters, equations developed for pseudoplastics can be applied to dilatant fluids. For example, the power law equation may be used with n greater than unity for dilatant fluids. Rheograms for these fluids are illustrated in Figure 1.2. Dilatancy is not as common as pseudoplasticity and is generally observed for fairly concentrated suspensions of irregular particles in liquids. A commonly accepted mechanism for dilatancy [refer to Metzner (1956)] is that at low shear rates the particles in such fluids are densely packed but still surrounded by liquid which lubricates the motion of adjacent particles. At higher shear rates, the dense packing breaks up, progressively forcing liquid out of more and more of the interstices between particles. There is now insufficient liquid to lubricate the motion of such adjacent "dried out" particles, and the shear stress increases with shear rate more quickly than before. Many dilatant fluids also exhibit volumetric dilatancy, that is, an increase in volume with shear rate, as well as viscous dilatancy, which is the increase in apparent viscosity with shear rate. Examples of dilatant fluids are aqueous suspensions of titanium oxide and suspensions of starch and quicksand.

Plastic Fluids (Fluids with Yield)

Certain fluids do not deform continuously unless a limiting or "yield" stress is exceeded. These are often referred to as plastic fluids, or *Bingham fluids,* although the latter implies that they follow a particular constitutive equation. In the simplest case, such fluids behave in a manner identical to Newtonian fluids once the yield stress is exceeded, in that their rheograms are Newtonian rheograms shifted upward [Figure 1.2(b)]. Pure Bingham behavior is rare in nature. In other cases, the flow curve is that of a pseudoplastic shifted upward, and such fluids are termed yield pseudoplastics. Yield dilatant behavior may also be encountered. The commonly accepted mechanism for the behavior of plastic fluids is that the fluid at rest contains a structure sufficiently rigid to resist shear stresses smaller than the yield stress τ_y. When this stress is exceeded, the structure collapses and deformation is continuous as for nonplastic fluids. Examples of fluids exhibiting plastic behavior are certain paints,

Table 1.6. Some Constitutive Equations for Pseudoplastic (or Dilatant) Fluids.

Model Name	Constitutive Equation	Apparent Viscosity	Limiting Newtonian Prediction Low Shear	Limiting Newtonian Prediction High Shear	Remarks[a]
Power Law (Ostwald-de Waele)	$\tau = K\dot{\gamma}^n$	$K\dot{\gamma}^{n-1}$	—	—	Two parameters: K, n. Newtonian limits not predicted.
Prandtl-Eyring	$\tau = A \sinh^{-1}(\dot{\gamma}/B)$	$A\dot{\gamma} \sinh^{-1}(\dot{\gamma}/B)$	A/B	—	Two parameters: A, B. Based on Eyring's kinetic theory of liquids.
Ellis	$\dot{\gamma} = (\phi_0 + \phi_1 \tau^{\alpha-1})\tau$	$(\phi_0 + \phi_1 \tau^{\alpha-1})^{-1}$	ϕ_0^{-1} for $\alpha > 1$ only	ϕ_0^{-1} for $\alpha < 1$ only	Three parameters: ϕ_0, ϕ_1, α. If $\phi_1 = 0$, gives Newtonian equation. If $\phi_0 = 0$, gives power law equation.
Reiner-Philippoff	$\dot{\gamma} = \left[\mu_\infty + \dfrac{\mu_0 - \mu_\infty}{1 + (\tau/\tau_s)^2}\right]^{-1} \tau$	$\mu_\infty + \dfrac{\mu_0 - \mu_\infty}{1 + (\tau/\tau_s)^2}$	μ_0	μ_∞	Three parameters: μ_0, μ_∞, τ_s. μ_0 and μ_∞ are the limiting viscosities.
Sisko	$\tau = a\dot{\gamma} + b\dot{\gamma}^c$	$a + b\dot{\gamma}^{c-1}$	a	—	Three parameters: a, b, c. Combination of Newtonian and power law.

[a] All these equations predict shear-dependent viscosity.

suspensions of finely divided minerals such as chalk in water, and some asphalts. The yield values can be quite small for water suspensions and very large for materials such as asphalt. Plastic behavior is necessary in paints to prevent flow when applied in the form of a vertical film. Constitutive equations for such fluids are modifications of the Newtonian or the power law expressions.

Bingham Model. The Bingham equation is given by

$$\tau = \tau_y + \eta\dot{\gamma}, \quad \tau > \tau_y$$
$$\dot{\gamma} = 0, \quad \tau < \tau_y \tag{1.39}$$

where τ_y is the yield stress, and η is the "plastic viscosity" or "coefficient of rigidity" (i.e., the slope of the flow curve in rectangular coordinates). On logarithmic coordinates, the curve is asymptotic to τ_y at low $\dot{\gamma}$ and approaches a slope of unity for high $\dot{\gamma}$ (Figure 1.2). Note that for $\tau < \tau_y$, $\dot{\gamma} = 0$, implying that there is no deformation in this region. The apparent viscosity for Bingham fluids is given by

$$\eta_a = \tau/\dot{\gamma} = \eta + \tau_y\dot{\gamma}^{-1} \tag{1.40}$$

showing that the apparent viscosity decreases with shear rate. For very high shear rates the effect of the yield stress becomes negligible, the apparent viscosity levels off to the plastic viscosity η, and the fluid exhibits Newtonian behavior.

Yield-Power Law Model. This model is expressed as

$$\tau = \tau_y + K\dot{\gamma}^n, \quad \tau > \tau_y$$
$$\dot{\gamma} = 0, \quad \tau < \tau_y \tag{1.41}$$

and is simply a combination of the Bingham and power law equations. As before, pseudoplastic or dilatant behavior develops depending on whether n is less than, or greater than, unity. An example of yield-pseudoplastic liquids is a clay-water suspension. Yield-dilatant behavior is less common.

Time-Dependent Fluids

Fluids described thus far manifest flow behavior that responds instantaneously to sudden changes in shear. For example, if a pseudoplastic fluid were subjected to a sudden change in shear rate, the new stress given by its constitutive equation would be achieved instantaneously. Such fluids are termed time-independent. Actually, a finite nonzero time is required for the new stress to be attained, but this time is very short. There is a class of fluids termed time-dependent fluids for which the response time is appreciable. For such fluids, the sudden application of a change in shear rate results in the shear stress changing slowly with time until a new equilibrium shear stress is established corresponding to the changed shear rate. The postulated mechanism for time-dependence is that, due to shear, the time scale for structural changes within the fluid is large compared to the time scale of shear and the very short time scales for time-independent fluids. As the flow behavior depends on the fluid structure, the shear stress responds slowly to an imposed change in shear rate. Thus, the shear stress becomes a function of shear rate and time until steady conditions are attained. Time-dependency may be exhibited by fluids that otherwise may be termed pseudoplastic, dilatant, or plastic. Time-dependent fluids are generally classified as either thixotropic or rheopectic.

Thixotropic fluids break down under shear. At a given shear rate, the shear stress slowly decreases until an equilibrium state is reached. Such fluids behave as time-dependent pseudoplastics. Thixotropic behavior is illustrated in Figure 1.3, which shows the change in apparent viscosity of the fluid when a higher shear rate $\dot{\gamma}_2$ is suddenly imposed at time t_1. At times less than t_1, the fluid has been sheared at $\dot{\gamma}_1$ for a long time so that an equilibrium apparent viscosity η_{a1} is established. For a time-independent pseudoplastic, the apparent viscosity decays to η_{a2} immediately. For a thixotropic fluid, a slow decay over some measurable time $(t_2 - t_1)$ occurs until the equilibrium viscosity η_{a2} is reached. If the new shear rate is a *decrease* over the original value $\dot{\gamma}_1$, the apparent viscosity will *increase* slowly with time. Such instantaneous changes in $\dot{\gamma}$ cannot be achieved in practice because of fluid and equipment inertia. A more practical approach to detecting thixotropy (or any

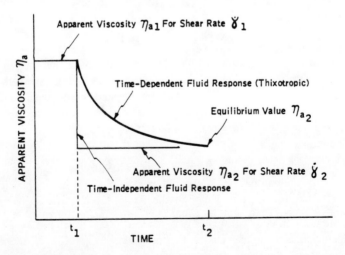

FIGURE 1.3. Response of time-dependent fluid to change in shear rate (thixotropic case).

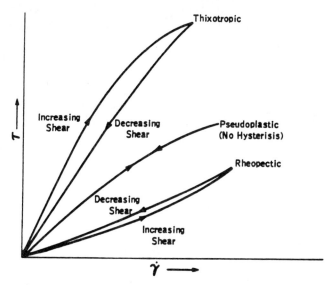

FIGURE 1.4. Hysteresis loops for time-dependent fluids (arrows show the chronology of the imposed shear rate).

time-dependent behavior) is to subject the fluid to a programmed change in shear rate with time, increasing from zero shear rate to some peak value and then decreasing back to zero. A thixotropic fluid under such a program would produce a hysteresis loop for τ versus $\dot\gamma$, as shown in Figure 1.4. For a pseudoplastic fluid no hysteresis is observed, and the same curve would be traced out for increasing and decreasing shear. Note that the area within the loop depends on the degree of thixotropy as well as the time scale of change of the shear rate. If the shear rate is changed slowly enough, even a highly thixotropic fluid produces a curve with no hysteresis. Also to be noted is the position of the curve for increasing shear relative to that for decreasing shear, which follows from the fact that the fluid is basically pseudoplastic, with an apparent viscosity that decreases with increasing shear rate. Examples of thixotropic fluids are paints, ketchup and food materials, oil well drilling muds and some crude oils. Thixotropy is necessary in paint to allow it to flow easily after the large shear imposed by brushing and then to recover a more viscous character after a short period of standing. Govier and Aziz (1977) have summarized some of the rheological models proposed for thixotropic fluids. An extensive review of the subject is given by Bauer and Collins (1967).

For *rheopectic fluids,* shear stress at a constant shear rate increases slowly with time until an equilibrium value is reached. Such fluids behave as time-dependent dilatant fluids. Under the programmed change in shear rate described above, the stress-shear rate curve forms a hysteresis loop, but of a different shape than a thixotropic fluid (Figure 1.4). As before, the location and shape of the loop depends on the shear rate program, as well as the degree of rheopexy. Rheopectic fluids are rare, examples being gypsum suspensions and bentonite clay suspensions.

Viscoelastic Fluids

If, for the fluids described, the imposed shear stress is removed, deformation ceases, but there is no tendency for the fluid to recover its original undeformed state. Certain fluids do have the property of partially recovering their original state after the stress is removed (i.e., they have memories). Such fluids thus have properties akin to elastic solids as well as viscous liquids, and are termed viscoelastic. Examples of viscoelastic fluids are molten polymers and polymer solutions, egg white, dough, and bitumens.

The elastic property of such fluids leads to some interesting and unusual behavior. The classic example is the phenomenon of rod-climbing, or the so-called "Weissenberg effect," exhibited by these fluids. If a rotating cylinder or rod is immersed in a purely viscous liquid, the liquid surface is depressed near the rod because of centrifugal forces. With a viscoelastic fluid, on the other hand, liquid climbs up the rod because of normal stresses generated by its elastic properties. This can be observed during mixing of flour dough and in stirred polymerization reactors. Another phenomenon is the marked swelling in a jet of viscoelastic fluid issuing from a die. As a result, extrusion dies must be designed properly to produce the desired product cross section.

The work performed on a viscoelastic fluid (for example, by forcing it through a tube) is stored in the fluid as normal stresses, as opposed to all being dissipated into heat in the case of purely viscous liquids. This stored energy is released when the fluid emerges from the tube and results in a swelling of the emerging fluid jet. The normal stresses generated within the tube relax when the fluid emerges from the tube; such fluids are said to exhibit *stress relaxation*. Some liquid-liquid mixtures consisting of droplets of one liquid dispersed in the other also exhibit viscoelasticity. Elastic energy is stored when the spherical droplets are distorted by shear and is released through the action of interfacial tension when the shear is removed.

Elastic effects are important mainly during the storage or release of elastic energy. Hence, such effects are important in the entrance and exit sections of tubes and during flow accelerations and decelerations caused by changes in cross section, by imposed oscillations, or by turbulence. For steady laminar flow in a tube or channel of constant cross section, elastic effects are not important except near the entrance and exit. On the other hand, for flow in fittings or in turbulent flow, these effects may become important. Pronounced viscoelasticity is observed mainly in molten polymers and concentrated polymer solutions and generally is not considered important for pipeline flow of other non-Newtonians such as slurries.

The flow behavior of these fluids cannot be represented by a simple relation between shear stress and shear rate alone. Instead, it will depend on the recent history of these quantities, as well as their current values. Constitutive equations for such fluids therefore must involve shear stress, shear rate, and their time derivatives. It is also clear that the time derivatives involved must be those applicable to a particular quantity of fluid as it moves through the system and not those from the viewpoint of a stationary observer, for example. The latter has no direct impact on fluid behavior; the former represents time rates of change experienced by the fluid itself. A large number of constitutive equations of this general type have been proposed involving, among other features, various types of time derivatives. Unfortunately, their use in engineering calculations is not widespread. For illustrative purposes, the Oldroyd model is noted, which is applicable for low shear rates:

$$\tau + \lambda_1 \frac{d\tau}{dt} = \mu \left(\dot{\gamma} + \lambda_2 \frac{d\dot{\gamma}}{dt} \right) \qquad (1.42)$$

Here, λ_1 and λ_2 are relaxation times and μ is the viscosity. We first observe that if λ_1 and λ_2 are both zero, the equation reverts to that for a Newtonian fluid. If only λ_2 is zero, the equation reduces to that for a Maxwell fluid. The Maxwell fluid is an early and simple model for a viscoelastic fluid based on a mechanical analog, whereby the fluid is represented as a spring and dashpot in series—the spring representing the elastic part and the dashpot the viscous part. Both the Maxwell fluid and the fluid represented by Equation (1.42) show stress relaxation. If flow is stopped,

$$\dot{\gamma} = \frac{d\dot{\gamma}}{dt} = 0$$

the stress decays or relaxes as e^{-t/λ_1}. If stress is removed, the shear rate in a Maxwell fluid becomes zero immediately, while for the fluid of Equation (1.42), the shear rate decays as e^{-t/λ_2}. This simple three-constant model has been shown to represent the behavior of certain viscoelastic fluids at low shear rates.

TYPES OF OPERATIONS AND APPROACHES TO PROBLEM ANALYSIS

Industrial operations may be generalized under three broad categories: intermittent or batch, semibatch, and continuous. An example of a batch operation is the production of paper pulp, in which wood chips, water and caustic cooking chemicals (called white liquor) are prepared in a reactor. The reactor vessel, called a digester, is charged with formulated proportions of these materials which are cooked in the presence of high-pressure steam. After a required cooking time, the digester is blown, i.e., steam and vapors are released, spent liquor is drained, and the cooked pulp is sent on to the next stage of processing. The digester is then ready for a new charge or batch of materials. Still another example of an intermittent operation is the production of beer and other alcoholic beverages in batch fermenter vessels.

An example of a semibatch operation is the production of chlorophenols by chlorination of phenols. The chlorophenols of greatest commercial importance are 2,4-dichlorophenol, an intermediate in the manufacture of 2,4-dichlorophenoxy-acetic acid (2,4-D) and its derivatives, which are selected herbicides, and pentachlorophenol (PCP), used as a wood preservative due to its fungicidal properties. In this semibatch process, phenol is charged into two batch reactors—a primary reactor and a secondary scrubber-reactor. Chlorine is added only to the primary reactor. The offgas from the primary reactor, consisting of chlorine and hydrogen chloride, is sent to the scrubber-reactor, where sufficient phenol is charged to ensure complete reaction of the chlorine. The hydrogen chloride offgas from the scrubber-reactor is recovered by dissolving it in water in an absorption tower to produce commercial-grade hydrochloric acid. The primary reactor is a stirred-batch reactor. A period of 8–10 hours is generally required for the chlorination [flowsheets and other examples are given by Goldfarb et al. (1981)].

Continuous operations and processes are the most desirable from both operating and cost standpoints. The process flowsheet of one example, the production of acetaldehyde by liquid-phase ethylene oxidation, is shown in Figure 1.5. Acetaldehyde is used primarily as an intermediate for the production of other organic chemicals—the major derivatives being acetic acid, acetic anhydride, n-butanol, and 2-ethylhexanol. Other products derived from acetaldehyde are pentaeythritol, trimethylol propane, pyridines, peracetic acid, crotonaldehyde, chloral, 1,3-butylene glycol, lactic acid, glyoxal, and alkylamines. In Figure 1.5, ethylene gas (stream 2), oxygen (stream 1), and a recycle gas stream (30) are continuously fed to a reactor containing an aqueous solution of palladium chloride and copper chloride. Exit stream (3) flows into a gas–liquid separator. The liquid stream from the separator (4) is split into two streams; stream (6) is recycled to the reactor and stream (5) is sent to a regenerator, where it is mixed with oxygen and steam to decompose copper oxalate and other organics prior to being returned to the reactor. The gas stream from the separator (11) contains the product (acetaldehyde), which is sent to a quench scrubber. There it is cooled and scrubbed with water to condense the acetaldehyde and other condensables. Noncondensed vapors (12) consisting of unreacted ethylene, oxygen, and various inerts are divided. Part of the stream (13) is purged to a flare to control the accumulation of inerts in the system, and the remainder (14) is recycled to the reactor vessel. Condensed stream (16) is split: stream (17) is recycled to the scrubber, and stream (18) is heated and fed to a light ends distillation column, in which dissolved gases and low boiling material are removed overhead (stream 19). The bottoms (20) from the light ends distillation column are fed to the product recovery column, where acetaldehyde is removed as an overhead stream (22). A by-product stream (27) is removed from the middle of the column. Water contaminated with residual organics (29) is removed from the bottom of the column and sent to disposal.

As illustrated by the above examples, process operations can be complex and comprise a large number of different unit operations. Although the nature of a particular process may be essentially intermittent, continuous production is often approached by the combination of a number of unit operations simultaneously but in different stages. Most chemical reaction processes are in essence semicontinuous when viewed as a whole operation; however, in a strict sense, they may be classified as batch when one is dealing with individual unit operations.

As illustrated by the elementary flowsheet for acetaldehyde production (Figure 1.5), each plant involves knowledge of an overwhelming accumulation of information on quantities and compositions of raw materials, intermediates, waste products, and by-products, which is essential to the production and accounting divisions and can be integrated into evaluating and improving an overall process. Quantitative bookkeeping of individual constituents is done by use of the generalized law of the conservation of mass. This is a general material balance applicable to all processes, with and without chemical reaction:

$$\left\{ \begin{array}{c} \text{Accumulation} \\ \text{of material} \\ \text{within the} \\ \text{system} \end{array} \right\} = \left\{ \begin{array}{c} \text{Input} \\ \text{through} \\ \text{system} \\ \text{boundaries} \end{array} \right\} - \left\{ \begin{array}{c} \text{Output} \\ \text{through} \\ \text{system} \\ \text{boundaries} \end{array} \right\}$$

$$+ \left\{ \begin{array}{c} \text{Material} \\ \text{accumulation} \\ \text{within the} \\ \text{system} \end{array} \right\} - \left\{ \begin{array}{c} \text{Consumption} \\ \text{within the} \\ \text{system} \end{array} \right\} \quad (1.43)$$

This relation may be applied to a single unit operation or to the entire process comprising many different unit operations. The same general balance may be applied to the energy relations of the system or to individual unit operations. In this form, another independent set of equations is formulated. The use of this principle is perhaps universal to engineering problems.

Most semicontinuous operations are repeated in an almost indefinite number of cycles, and both continuous and semicontinuous operations are

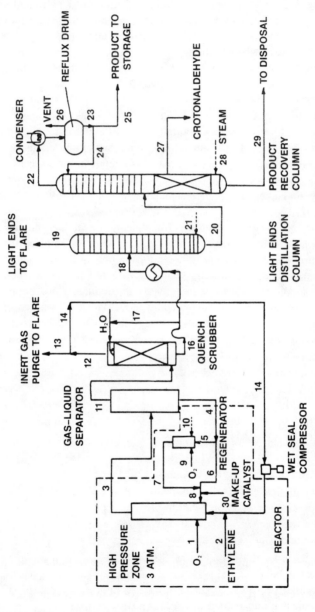

FIGURE 1.5. The continuous single-stage process for manufacturing acetaldehyde from ethylene.

performed over relatively long time periods. As such, the accumulation of materials and/or energy in a system or piece of equipment is limited. If the operations are performed over a sufficient time such that fluctuations of amounts are well known and negligible in comparison to input and output streams, the last two terms in Equation (1.43) may be neglected, whence we may state simply that

$$(\text{Total input}) = (\text{output}) + (\text{losses}) \qquad (1.44)$$

This simple relation may be applied (1) to evaluating the performance of existing equipment and operations; and (2) in the design of equipment and planning new operations. This book is primarily concerned with the latter. Forthcoming chapters will apply this concept and some of the others briefly described in this introductory chapter to the design and proper selection of unit operations concerned with the physical combination and separation of process streams.

We close this chapter with some general guidelines to problem analysis and solving. The success of a solution to any engineering problem is directly proportional to the degree of understanding of the mechanisms responsible for a particular phenomenon. For example, if our concern is with the production of a chemical feedstock, an understanding of the basic chemistry alone is insufficient. The environment in which the reaction is to take place is of equal concern. How feedstreams are to be combined and in what form in a commercial operation are questions addressed by both process kinetics and principles of unit operations. The emphasis in this text is on the hydrodynamics. Without this information the rational design of any process is simply not possible.

In the solution of problems, the following steps will be helpful:

1 Prepare a sketch of the phenomenon or process. This may be as detailed a drawing as needed to completely understand the requirements of the problem or the important mechanisms.

2 If a mathematical model is to be developed describing the phenomenon or process, or an experimental unit is to be constructed, list all important mechanisms. From this list, decide which ones are the most important and which may be neglected in the analysis.

3 Select a basis for the design calculations. For example, if a piece of equipment is being scaled-up from a cold model unit, the basis to ensure proper operation might be residence time or an appropriate dimensionless group related to operating performance.

4 Tabulate all data and intermediate steps in design computations. In this day of computers, this last step is self-evident.

NOTATION

A = area, m²
a = acceleration, m/s²
C = molal heat capacity, cal/(g)(°C)
F = area, m²
F_0 = force, N
g = local acceleration due to gravity, m/s²
g_c = conversion factor, 32.174 lb$_m$-ft/lb$_f$-s²
H = enthalpy, J/mole
K = consistency index
KE = kinetic energy, J
L = length, cm
M = mass, kg
m = mass, kg
n = number of moles for flow behavior index
p = pressure, Pa
PE = potential energy, J
Q = heat
R = ideal gas law constant
S = distance, m
T = absolute temperature, K or °R
t = time, s, or temperature, °C or °F
U = internal energy, J
u = velocity, m/s
V = volume, m³
v = velocity, m/s
W = work, N-m
Z = elevation, m

Greek Symbols

γ = ratio of specific heats, $c_p/c\dot{v}$
$\dot{\gamma}$ = shear rate (rate of deformation), s⁻¹
η_a = apparent viscosity, poise
θ = time, s
λ = relaxation time, s
μ = viscosity, poise
ϱ = density, kg/m²
τ = stress, Pa
τ_y = yield stress, Pa
$\dot{v}$ = molal volume, m³/mol

REFERENCES

BAUER, W. H. and E. A. Collins. *Rheology, Vol. 4*. F. R. Eirich, ed. New York:Academic Press Inc., pp. 423-459 (1967).

BIRD, R. B., R. C. Armstrong and O. Hassager. *Dynamics of Polymeric Liquids, Vols. I & II*. New York:John Wiley & Sons, Inc. (1977).

BIRD, R. B., W. E. Stewart and E. N. Lightfoot. *Transport Phenomena*. New York:John Wiley & Sons, Inc. (1960).

CHEREMISINOFF, N. P. *Fluid Flow: Pumps, Pipes and Channels*, Ann Arbor, MI:Ann Arbor Science Publishers (1981).

Federal Register, 42(206)56513-56514 (October 26, 1977).

FREDRICKSON, A. G. *Principles and Applications of Rheology*. Englewood Cliffs, NJ:Prentice Hall, Inc. (1964).

GOLDFARB, A. S., G. R. Godgraben, E. C. Herrick, R. P. Ouellette and P. N. Cheremisinoff. *Organic Chemicals Manufacturing Hazards*. Ann Arbor, MI:Ann Arbor Science Publishers (1981).

GOVIER, G. W. and K. A. Aziz. *The Flow of Complex Mixtures in Pipes*. New York:Kreiger Publishing Co. (1977).

LODGE, A. S. *Elastic Liquids*. New York:Academic Press, Inc. (1964).

METZNER, A. B. *Advances in Chemical Engineering, Vol. 1*. T. B. Drew and J. W. Hoopes, ed. pp. 79-150 (1956).

MIDDLEMAN, S. *The Flow of High Polymers: Continuum and Molecular Rheology*. New York:Interscience Publishers (1968).

SCHOWALTER, W. T. *Mechanics of Non-Newtonian Fluids*. Oxford, England:Pergamon Press, Inc. (1978).

SKELLAND, A. H. P. *Non-Newtonian Flow and Heat Transfer*. New York:John Wiley & Sons, Inc. (1967).

SMITH, J. M. and H. C. Van Ness. *Introduction to Chemical Engineering Thermodynamics*, 2nd ed., New York:McGraw-Hill Book Co. (1959).

SRIDHAR, T. "Transport Properties of Liquids," in *Handbook of Fluids in Motion*. N. P. Cheremisinoff and R. Gupta, eds. Ann Arbor, MI:Ann Arbor Science Publishers, pp. 3-27 (1983).

CHAPTER 2

Similarity, Modeling, and Dimensional Analysis

INTRODUCTION

Often, problems of a unique nature are encountered in industry. Therefore, many processes cannot be formulated rigorously in a purely mathematical sense. Physical and chemical phenomena may be too complex to be described from existing theories. To formulate mathematical relationships that characterize a process and are applicable to design, experimental investigations are conducted. From experimental observation and data, empirical equations can be developed. However, these correlations are often local in character and their range of applicability is limited to the conditions over which studies were performed. Despite these shortcomings, such correlations have value and are used, perhaps too often, in designing and scaling up unit operations in chemical engineering.

The value of experiments can be enhanced greatly if results are generalized. In this manner, those parameters that best characterize the behavior of a process can be identified. By utilizing principles of *similarity theory*, experimental observations from small-scale tests may be applied with greater confidence to designing and predicting the performance of large-scale equipment.

Similarity theory is a method that provides a scientific approach to generalizing experimental results. It provides a means of evaluating important process parameters, and the technique can be used to limit the number of costly experiments needed to obtain unified equations. That is, it permits studies of the performance of the system (prototype) to be based on a small-scale replica (referred to as a "model").

The term *model* in scientific literature does not always refer to a prototype's replica in which experiments can be conducted. It most generally refers to a cognitive or conceived physical mathematical description, i.e., a scheme that attempts to describe the most important variables of a process. In this chapter, however, the term will refer to physical and material approximations of the true system under investigation. Through physical modeling, the nature and behavior of a process may be investigated not only on a smaller scale than the conceptual design, but with different substances and at different temperatures, pressures, etc. In this manner, generalized dependencies can be defined mathematically. This allows experiments to be performed under less severe conditions than in the commercial case.

Similarity theory and physical modeling allow us to investigate phenomena more rapidly and economically, with a reasonable degree of reliability in making the transition from laboratory scale to commercial scale. Note, however, that they cannot provide complete theoretical understanding of the physicochemical phenomena in terms of fundamental equations. The theory and practice only provide integral solutions of the theoretical equations which define the process by generalizing experimental data. These data are valid only for a group of similar phenomena in the range of investigation and have the advantage of negating formal investigation of the theoretical eqautions.

PRINCIPLES OF SIMILARITY THEORY

Similarity theory is based on the principles of selecting a group of analogous phenomena from unrelated processes. For example, although the motion of fluids in the atmosphere and through piping are different systems, these flow phenomena are analogous because both represent the motion of viscous fluids under the influence of pressure gradients. Hence, the fluid motion for these different systems may be described by the unified Navier-Stokes equations and thus considered to be of the same class of phenomena. Similarly, the motion of incompressible and compressible viscous fluids through piping and equipment, although considerably different processes, comprise a class of similar phenomena. Phenomena are considered operationally similar when the ratios of certain parameters characterizing the system are compatible.

First, we shall consider the condition used most often to determine operational similarity, namely the *geometric* condition. For a model to be geometrically similar to the prototype, the ratio of the distances between any two common points in both systems must be constant. This ratio is referred to as the *geometric scale factor*. For example, the characteristic dimensions of a drum dryer are its diameter D and length L. The geometric scale factor is thus L/D, and the model will be similar to the prototype when the following condition is fulfilled:

$$\Lambda = \frac{L_1}{D_1} = \frac{L_2}{D_2} \tag{2.1}$$

where L and D have the same units (m) and, thus, the ratio is dimensionless. Subscripts 1 and 2 refer to the model dryer and prototype, respectively. Therefore, the model's diameter differs from the prototype by some constant scale factor. Thus, similar ratios can be defined for any geometric configuration.

Parameters of the same generic class are interchangeable; that is, those parameters that determine the similarity scale factors may be changed by similar values, which define the system geometry. Hence,

$$\frac{\ell'}{\ell''} = \frac{\ell'_1}{\ell''_1} = \frac{\ell'_2}{\ell''_2} = \frac{\ell'_1 - \ell'_2}{\ell''_1 - \ell''_2} = \frac{d\ell'}{d\ell''} = \Lambda_\ell \qquad (2.2)$$

Geometric similarity between systems is a necessary condition for similarity of physical phenomena but is not sufficient in itself. For physical similarity to exist, all important parameters influencing the phenomena must be similar. These parameters often change as functions of time and space in each system. Hence, technological processes are similar only under conditions of mutual fulfillment of geometrical and time similarities over the fields of physical values, as well as similarities of initial and boundary conditions. Similarity among physical parameters such as velocity, acceleration, density, pressure, etc., may be defined in a manner analogous to geometric conditions:

$$\left.\begin{array}{l} \text{For velocity, } v_1 = \Lambda_v v_2 \\ \text{For acceleration, } a = \Lambda_a a_2 \\ \text{For density, } \varrho_1 = \Lambda_\rho \varrho_2 \\ \text{For pressure, } P_1 = \Lambda_p P_2 \\ \text{For time, } t_1 = \Lambda_t t_2 \end{array}\right\} \qquad (2.3)$$

The similarity scale factors $\Lambda_\ell, \Lambda_v, \Lambda_\varrho, \Lambda_p \ldots$ are constant for different compatible points of two similar systems but change depending on the size ratio of the prototype to the model.

Process similarity between the prototype and the model may be determined through the use of *characteristic parameters*. These parameters are expressed in the form of compatible ratios in the limit of each system and are illustrated in Figure 2.1. We shall denote a "dummy" characteristic parameter by i.

$$i = \frac{\ell'_1}{L'} = \frac{\ell''_1}{L''} = \text{inv.} \qquad (2.4)$$

where inv. denotes invariantly or "one and the same."

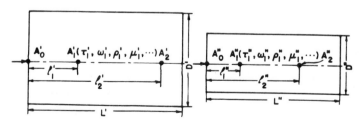

FIGURE 2.1. Scheme for the formulation of similarity conditions.

A parameter may be expressed in terms of relative units and, hence, an appropriate scale can be defined. For example, the diameter of the dryer may be selected as a scale rather than a length. For other systems, it may be appropriate to define this scale relative to common points in the model and prototype. Hence, from Equations (2.3) and (2.4), we write the following expressions in which the measuring scale is defined relative to the entrance of the system (i.e., in terms of the initial conditions v'_0, v''_8, etc.):

$$\left.\begin{array}{c} \dfrac{t'_1}{T'} = \dfrac{t''_1}{T''} = i_t \\[1em] \dfrac{v'_1}{v'_0} = \dfrac{v''_1}{v''_0} = i_v \\[1em] \dfrac{\varrho'_1}{\varrho'_0} = \dfrac{\varrho''_1}{\varrho''_0} = i_\varrho \\[1em] \dfrac{P'_1}{P'_0} = \dfrac{P''_1}{P''_0} = i_p \end{array}\right\} \qquad (2.5)$$

The characteristic parameters, i_t, i_v, i_ϱ, i_p . . . (referred to as simprexes), may not be equal to each other for different compatible points of similar systems; they are independent of the size ratio of the prototype and model.

This means that in passing from one system to the other the characteristic parameters do not change. For example, consider a dryer of total length $L = 10$m and assume a portion of fluid flows from a section I-I (a distance of 2m from the inlet of the dryer) to the section II-II (a distance of 5m from the inlet. The characteristic parameter changes from $i_1 = 2/10 = 0.2$ to $i_2 = 5/10 = 0.5$; however, the scale factor remains unchanged if the sizes of the prototype and model are kept constant.

Let us assume the gas velocity in the dryer entrance to be $v'_0 = 4$ m/s and, at the middle section of the dryer, $v' = 3$ m/s. Consequently, the characteristic parameter $i_v = v'/v_0 = 3/4 = 0.75$. To provide similarity of the velocity field, it is necessary that $v' = v'' = 0.75$ for compatible points of the model. If the entrance velocity in the model $v''_0 = 3$ m/s, then at the middle of the model the velocity scale is

$$v'' = v''_0 \times v' = 3 \times 0.75 = 2.25 \text{ m/s}$$

METHODS OF SIMILARITY THEORY

The mathematical principles in similarity theory are simple and easily implemented. However, the student is cautioned against adopting a formal or stereotyped approach to avoid direct errors that can result if the physical concepts of the methods are not adopted. The application of specific techniques or theorems is determined by a volume of preliminary knowledge about the process under consideration. Application of the theory to the development of fully integrated equations is known as dimensional analysis. There are three basic theorems, each of which addresses a separate issue concerning the planning of model experiments. These issues are:

1 Identification of important parameters to be measured in the experiments
2 The final and most useful form in which the experimental results should appear
3 The type of equipment to which the model's experimental results may be applied

Newton's Theorem

Newton's theorem addresses the first issue. It is formulated in two ways:

1 According to Newton, *the conditions necessary (and sufficient) for similarity of the phenomena are equality of the values of the dimensionless groups made up of the quantities given in the conditions.*
2 According to Kirpichev (1953) [see also Gukhman (1965) and Sedov (1959)], *the conditions necessary (and sufficient) for similarity of the phenomena are equality of characteristic parameters (similarity indicators) to unity.*

We will illustrate the correctness of these formulations in the following example. Consider two similar systems that satisfy Newton's second law (the momentum equation), determining the relation between external forces and produced acceleration.

From Newton's law, the total force acting on a body is

$$f = m \frac{dv}{dt} \qquad (2.6A)$$

where

f = force (N)
m = mass (kg)
v = velocity (m/s)
t = time (s)

For two similar systems, Equation (2.6A) may be written twice:

$$f_1 = m_1 \frac{dv_1}{dt_1} \tag{2.6B}$$

$$f_2 = m_2 \frac{dv_2}{dt_2} \tag{2.6C}$$

Physical values for the two systems differ only by a scale factor; therefore,

$$f_1 = \Lambda_f f_2; \ m_1 = \Lambda_m m_2; \ v_1 = \Lambda_v v_2; \ t_1 = \Lambda_t t_2;$$

$$dv_1 = \Lambda_v dv_2; \ dt = \Lambda_t dt_2$$

Dividing Equation (2.6B) by Equation (2.6C) we obtain

$$\frac{f_1}{f_2} = \frac{m_1}{m_2} \frac{dv_1}{dv_2} \frac{dt_2}{dt_1}$$

or

$$\Lambda_f = \Lambda_m \frac{\Lambda_v}{\Lambda_t} \tag{2.7}$$

Consequently, we obtain

$$\frac{\Lambda_f \Lambda_t}{\Lambda_m \Lambda_v} = 1 = \tilde{j} \tag{2.8}$$

According to Kirpichev's statement, the nondimensional parameter $\tilde{j}$ (indicator of similar transformation) for two similar phenomena is equal to unity. Therefore, the selection of numerical values of scale factors is not arbitrary, but rather subjected to the conditions of $\tilde{j} = 1$.

As follows from Equation (2.7)

$$\frac{f_1 dt_1}{m_1 dv_1} = \frac{f_2 dt_2}{m_2 dv_2}$$

and

$$\frac{f_1 t_1}{m_1 v_1} = \frac{f_2 t_2}{m_2 v_2} \tag{2.9}$$

This last expression is a dimensionless group known as the Newton number:

$$Ne = \frac{ft}{mv} \tag{2.10}$$

or, taking into account that $t = \ell/v$,

$$Ne \equiv \frac{f\ell}{mv^2} \quad (2.11)$$

The symbol "≡" indicates that this is a definition and does not denote the Newton number to be a function of the f, t, m, v, i.e., $Ne \neq f(ft/mv)$. Thus, for a series (groups) of similar processes whose class is described by initial physical equations based on Newton's second law, the following equality is correct:

$$Ne_1 = Ne_2 = Ne_3 = \ldots = \text{idem}$$

Hence, if we consider similar processes of motion in a model and in a prototype, then

$$Ne_{model} = NE_{prototype} \quad (2.12)$$

Equation (2.12) is the mathematical formulation of the first theorem of similarity. Thus, with properly planned experiments it is only necessary to measure those values that appear in the dimensionless groups of a process under evaluation. The dimensionless groups constitute generalized characteristics of a process consisting of dimensional physical values reflecting different characteristics of a phenomenon.

Because the dimensions entering these groups are reduced, they have a zero dimension, and the numerical value of a dimensional group remains true whatever the system of units in which the various quantities are measured.

To derive and compute dimensionless groups properly, we must be sure that the initial equation or expression is dimensionally homogeneous, i.e., that the dimensions of all the terms in each group are consistent and cancel. The correctness of deriving dimensionless groups is checked by reducing the dimensions from which the group is formed. In the case considered,

$$Ne = \frac{ft}{mv}\left[\frac{N-s^2}{kg-m} = \frac{N}{N}\right]$$

i.e., the dimensions are reduced and, consequently, the dimensionless group is correct. The *dimensionless groups* are derived from the *dimensional* equation governing the physical process. The essential premise of deriving these groups is the availability of the equation governing the process, i.e., its mathematical description. Regardless of whether this equation is expressed in algebraic or differential form, the dimensionless groups may be derived by the same technique from any homogeneous dimensional equation.

Dimensionless groups can be formed simply by dividing through the equation by any dimensional product that makes all the terms of the equation dimensionless. Thus, they can be formed directly from the basic equation of the system and *without its formal solution*.

In addition to the basic physical equation, the derived form of the dimensionless group has an intrinsic physical character. For example, the Newton number expresses the ratio of active to reactive forces and is therefore a measure of impulse, *ft*, and momentum, *mv*. In special cases of motion depending on concrete expressions for force, mass, and velocity, the Newton number can take on another form; that is, it can be transformed into the Reynolds, Euler, Froude, or Stokes numbers.

To derive these dimensionless numbers, it is necessary to substitute the acting forces into the expression of the Newton number accordingly. For the *Froude number*,

$$Ne_1 = \frac{[mg]\ell}{mv^2} = \frac{g\ell}{v^2}$$

$$Fr = \frac{v^2}{g\ell}$$

(2.13)

Fr is the ratio of inertia force on a fluid element to the gravity force.

For deriving the *Reynolds number* from the Newton number, we write down the expression of the friction force in the flow of a viscous fluid as it is determined by Newton's law:

$$f = \mu \ell^2 \frac{dv}{d\ell}$$

(2.14)

Substituting *f* from Equation (2.14) into Equation (2.11) and denoting $m = \varrho \ell^3$ (*m* is related to a unit volume), we obtain

$$Ne_2 = \frac{\left[\mu \ell^2 \frac{dv}{d\ell}\right]\ell}{[\varrho \ell^3]v^2} \; ; \; Ne_2 d\ell = \frac{\mu dv}{\varrho v^2}$$

$$Ne_2 \int_0^\ell d\ell = \frac{\mu}{\varrho} \int_0^{v_{max}} v^{-2} dv; \; Ne_2 = \frac{\mu}{\varrho v \ell}$$

or

$$Re = \frac{dv\varrho}{\mu} \qquad (2.15)$$

or

$$Re = \frac{dv\gamma}{\mu g} \qquad (2.15A)$$

The Reynolds number is interpreted as the ratio of inertial forces to the viscous forces in the flow.

The *Euler number*, which characterizes the hydrodynamic processes running under action of mechanical pressure, is the ratio of static pressure drop, Δp, to dynamic head, ϱv^2:

$$Ne_3 = \frac{[\Delta p \ell^2]\ell}{[\varrho \ell^3]v^2} \; ; \; Eu = \frac{\Delta p}{\varrho v^2} \qquad (2.16)$$

The *Stokes number* is important in analyzing sedimentation processes. Substituting the resistance force of the medium into the Newton number,

$$R = 3\pi d\mu v_1 \qquad (2.17)$$

$$Ne_4 = \frac{[3\pi d\mu v]\ell}{\left[\frac{\pi d^3}{6}(\varrho_1 - \varrho_2)\right]v_2^2} = \frac{\mu \ell}{d^2 \varrho_1 v^2} \qquad (2.18A)$$

or

$$Stk = \frac{d^2 \varrho_1 v_2}{\mu \ell} \qquad (2.18B)$$

The Stokes number is the ratio of the resistance force of the medium to the buoyancy force of a particle. The intrinsic physical character of each dimensionless number (criterion of similarity) differs from *arbitrarily chosen* dimensionless complexes composed of random physical values.

From the equations governing the physical processes of different classes, we obtain dimensionless numbers (criteria of similarity) of different processes: thermal, diffusion, hydrodynamic, and so on. The more complicated equations and systems of equations give at once *several dimensionless numbers,* describing different characteristics of a complicated process. The technique described above for deriving dimensionless numbers is not unique. Further discussions are given by Sedov (1959), Gukhman (1965), Ipsen (1960), Langhaar (1951), and Lighthill (1963). Other methods are illustrated below.

Derivation of Dimensionless Groups from Process-Governing Equations

There are three basic methods for deriving dimensionless groups from a governing equation describing the process. Each is illustrated by considering the steady accelerated motion of a body:

$$w = w_0 + at \qquad (2.19A)$$

where

w = velocity at time t (m/s)
w_0 = velocity at time $t = 0$ (m/s)
a = acceleration (m/s^2)
t = time from starting motion (s)

Method I

This method involves variable transformation through the use of scale factors employed earlier in deriving the Newton number.

For the first phenomenon,

$$w_1 = w_{01} + a_1 t_1 \qquad (2.19B)$$

For the second phenomenon,

$$w_2 = w_{02} + a_2 t_2 \qquad (2.19C)$$

For the similarity of two phenomena, we have

$$\frac{w_1}{w_2} = C_w; \; w_1 = C_w w_2; \; \frac{w_{01}}{w_{02}} = c_2; \; w_{01} = c_w w_{02}$$

$$\frac{a_1}{a_2} = C_a; \; a_1 = C_a a_2; \; \frac{t_1}{t_2} = C_t; \; t_1 = C_t t_2$$

Substituting the new notations of w_1, w_{01}, a_1, and t_1 into the expression for the first phenomenon (Equation 2.19B), we obtain

$$C_w w_2 = c_w w_{02} + C_a C_t a_2 t_2 \qquad (2.20)$$

Equations (2.19B) and (2.20) can coexist only under the condition of reduction of multiples formed from factors. This is equivalent to the condition of equality in pairs:

$$C_w = c_w; \quad C_w = C_a C_t$$

The last condition gives two *characteristic parameters* (indicators of similarity):

$$\tilde{j}' = \frac{C_w}{c_w} = 1$$

$$\tilde{j}'' = \frac{C_a C_t}{C_w} = 1$$

In the expression for $\tilde{j}'$, C_w is related to *different* process velocities, but it cannot be reduced.

Substituting the scale factors by the ratios of values, we obtain

$$\tilde{j}' = \frac{\dfrac{w_1}{w_2}}{\dfrac{w_{01}}{w_{02}}} = 1; \quad \frac{w_1}{w_{01}} = \frac{w_2}{w_{02}} = \text{idem}$$

Hence, the first dimensionless number, a *dimensionless velocity* (in this case a relative velocity) is

$$K_1 \equiv \frac{w}{w_0} \qquad (2.21)$$

In the same manner we obtain from the expression $\tilde{j}''$ the second dimensionless number:

$$\tilde{j}'' = \frac{\dfrac{a_1}{a_2} \times \dfrac{t_1}{t_2}}{\dfrac{w_{01}}{w_{02}}} = 1; \quad \frac{a_1 t_1}{w_{01}} = \frac{a_2 t_2}{w_{02}} = \text{idem}$$

Hence, the second dimensionless number of time similarity (i.e., the criterion of kinematic homochronicity) is

$$K_2 \equiv \frac{at}{w_0} \qquad (2.22)$$

A constant-scale value w_0 is assumed in both expressions for the dimensionless numbers. According to the terms of the initial equation, we have to determine

$$\frac{w}{w_0} = f\left(\frac{at}{w_0}\right)$$

or

$$K_1 = f(K_2)$$

(2.23)

In this simple case, the form of this function is understood in the second method of similarity transformation.

Method II

This method consists of dividing all terms of the homogeneous equation by one of its terms serving as a scale (in this case by w_0):

$$\frac{w}{w_0} = 1 + \frac{at}{w_0}$$

Hence,

$$K_1 = \frac{w}{w_0} \; ; \; K_2 = \frac{at}{w_0}$$

whence the form of the function is evident:

$$K_1 = 1 + K_2 \tag{2.24}$$

The equations governing the process are usually much more complicated, and the form of the function is not so readily determined. However, for clarification of the combination of physical values entering into the dimensionless numbers, the method is applicable.

Method III

This method consists of transition to new independent units of measurement of physical values. Assume we know a priori that there is an explicit relationship:

$$w = f(w_0, a, t) \tag{2.25}$$

Here, the base units of measure are m and s, and the dependent units of measure are m/s (for velocity) and m/s² (for acceleration). Now we transfer to new independent units of measure which are less in L and T times than the first ones. Then the numerical values of w, a, and t will be changed to

$$w \frac{L}{T} = f\left(w_0 \frac{L}{T}, a \frac{L}{T^2}, tT\right) \qquad (2.26)$$

Fluid motion is independent of the units chosen for measuring specific process characteristics. Therefore, the basic Equation (2.25) has to retain its structure at different values of coefficients L and T. The numerical values of these coefficients are chosen to provide expressions that are simple and convenient in application. In particular, we may choose L and T as

$$w_0 \frac{L}{T} = 1; \ a \frac{L}{T^2} = 1; \ tT = 1$$

Then,

$$T = \frac{1}{t}; \ \frac{L}{T} = \frac{1}{w_0}; \ \frac{L}{T^2} = \frac{1}{a}; \ \frac{1}{w_0} = \frac{T}{a} = \frac{1}{at}$$

$$a \frac{L}{T^2} = a \frac{1}{w_0} \times \frac{1}{T} = \frac{at}{w_0}$$

Substituting the obtained values into Equation (2.26), we have

$$\frac{w}{w_0} = f\left(1, \frac{at}{w_0}, 1\right)$$

or

$$\frac{w}{w_0} = f\left(\frac{at}{w_0}\right)$$

or

$$K_1 = f(K_2)$$

This coincides with the result of transformation by the first method.

All these methods are applicable if one has the equation governing the physical process, i.e., the mathematical description of the process containing all necessary characteristic physical values.

However, as is more often the case, only the *qualitative* description of a process is known; the quantitative relationships among the different physical factors and their forms are unknown. Then the theory of similarity helps to determine the *hypothetical* form and prioritizes the most important dimensionless numbers of the process. We may then use carefully planned experiments to validate this hypothesis. As noted earlier, the method used to predict the form of a dimensionless number without knowledge of the basic equation is called dimensional analysis. Its application is conjugated with Buckingham's "pi" theorem.

Dimensional Analysis

We shall illustrate this method on the same example by assuming that the governing equation of the process [Equation (2.19)] is unknown. Hence, the only information that is known is

$$w = f(w_0, a, t)$$

Denoting the dimensions of the various parameters symbolically:

$$w[\text{m/s}] = LT^{-1}$$
$$w_0[\text{m/s}] = LT^{-1}$$
$$a[\text{m/s}^2] = LT^{-2}$$
$$t[\text{s}] = T$$

The dimensional equation describing the process may be assumed to have the form of a power function:

$$w = C w_0^x a^y t^z \tag{2.27}$$

where C = constant of the equation determined from experiments, and x, y, z are powers to be determined. Instead of Equation (2.27), we write the equation of dimensions in accordance with the assigned symbols:

$$LT^{-1} = C(LT^{-1})^x(LT^{-2})^y T^z$$

or

$$LT^{-1} = CL^{x+z}T^{-x-2y+z} \tag{2.28}$$

The powers of the symbols on both sides of the expression must be equal:

$$1 = x + y$$

$$-1 = x - 2y + z$$

Thus, we have two equations with three unknowns, which may be solved by inspection. Assume $z = +1$; then $x = 0$ and $y = 1$. Consequently,

$$w = Cw_0^0 at = Cat$$

The dimensionless number $K_1 = w/w_0$ and w_0 result from the examination. Only one dimensionless number may be derived from the remaining physical values $-K \equiv at/w$, with a *variable* scale w, which is undesirable in this case. Therefore, we will consider the second variant of Equation (2.27):

$$\frac{w}{w_0} = Ca^x t^y \qquad (2.29)$$

For this equation, the formula of dimensions is

$$L^0 T^0 = C(LT^{-2})^x T^y$$

And comparing the powers, we obtain

$$0 = x; \; 0 = -2x + y; \; y = 0$$

These conditions are possible at zero dimension of the right-hand side (RHS) of the equation, which is achieved by dividing through by w_0:

$$\left[\frac{LT^{-2}T}{LT^{-1}}\right] = [LT]^0$$

Therefore,

$$\frac{w}{w_0} = C\left(\frac{at}{w_0}\right)$$

or

$$K_1 = f(K_2)$$

The form of this function may be determined experimentally.

As shown, the application of dimensional analysis involves the logical selection of values and may give a positive result only if the initial set of all factors governing the process is selected properly. However, it is not always possible for a process to be analyzed thoroughly without experiments. Therefore, we cannot exclude the possibility of missing a dimensionless number, especially from the numbers having the same name, such as $K_1 \equiv w/w_0$ (simplex number, differing from a complex number such as $K_2 \equiv at/w_0$ formed from different values). Therefore, dimensional analysis may be used in simple cases of generalization where there is not a large number of variables.

To restore the "lost" numbers, Buckingham's "pi" theorem may be employed with all methods of deriving dimensionless numbers. Further discussions of the above analysis are given by Klinkenberg (1955).

Buckingham Pi Theorem

The Buckingham pi theorem is a rule for determining the number of dimensionless groups that exist [called π's, in Buckingham's (1914) notation].

The theorem states that any dimensionally homogeneous equation connecting N physical values, the dimensions of which are expressed by n fundamental units, can be reduced to a functional relationship between π dimensionless numbers:

$$\pi = N - n \qquad (2.30)$$

The number of dimensionless groups (called simplexes) containing π numbers is equal to the number of pairs of the same values in the basic equation. Equation (2.30) states the rule for determining the number of dimensionless groups characteristic of the process. For our example, with the basic equation $w = w_0 + at$,

$$N = 4(w, w_0, a, t); \quad n = 2(\text{m/s})$$

Consequently, the total number of $\pi = N - n = 2$ (where $K_1 \equiv w/w_0$; $K_2 \equiv at/w_0$). Among them, one dimensionless number is a simplex $K_1 \equiv w/w_0$ because the basic equation contains one pair of the same values w and w_0.

Thus, using the first theorem of similarity and its associated methods, we obtain a definite amount of dimensionless numbers, i.e., generalized process characteristics for any basic physical equation. Among these could be simplex dimensionless numbers (so-called *parametric* dimensionless numbers of a geometric or physical nature, which are relative sizes of a system reflecting similarity) and complexes (called *numbers,* such as the Reynolds number, Froude number, Euler number, etc.). Any combination of the laws of similarity (dimensionless numbers) is also a criterion of similarity. The numbers may be divided, multiplied, and/or raised to powers by each other to obtain new dimensionless groups. The transition from dimensional characteristics of a process to generalized characteristics decreases considerably the amount of variables. Further simplification may be achieved by combining dimensionless groups.

Lighthill (1963) notes that one of the most significant consequences of Buckingham's theorem is the economy it makes possible in the number of variables that need to be included in an experimental investigation or theoretical analysis. The number of independent dimensionless products, whose values determine the properties of the system, is generally less than the number of different kinds of physical quantities in the system. The consequent reduction in the number of experimental conditions that need to be covered in a program of experiments represents an enormous simplification and economy, resulting in simplifications in the plotting or tabulation of results. Similar simplifications are achieved in analytical solutions when problems are formulated in terms of dimensionless products at the outset.

Federman-Buckingham's Theorem

Federman-Buckingham's theorem [see Gukhman (1965) and Kirpichev (1953) for details] constitutes the second theorem of similarity. It states that the quantitative results of experiments should be presented by equations expressing the relationship among nondimensional numbers. The dimensionless groups K_1 containing the parameters of interest should be expressed as a function of other dimensionless numbers reflecting different sides of a process:

$$K_1 = f(K_2, K_3, K_4 \ldots) \tag{2.31}$$

In the examples given above, the form of the function was known, see Equation (2.24). Generally, the form of this function is not known a priori and must be determined from experimental data. Unlike human beings, nature rigorously applies the law of geometric progression and probability. Man attempts to approximate these laws through logarithmic relationships. The results of experimental investigations are approximated more readily in either form:

$$K_1 = CK_2^m K_3^n K_4^p \tag{2.32}$$

or exponential forms for kinetic processes:

$$K_1 = e^{-\tau/\theta} \text{ (damping process)} \tag{2.33}$$

$$K_1 = K_0(1 - e^{-t/\theta}) \text{ (growing process)} \tag{2.34}$$

where C, m, n, and p are constants determined from the graphic analysis of experimental data.

K_0 is the initial value of the dimensionless number K_1 at time $t = 0$, and t denotes the time from the beginning of the process. θ is the time constant of the process that depends on the conditions of its realization and is expressed through the dimensionless numbers.

The second theorem of similarity is formally stated as follows. *The solution of any differential equation may be presented as a relationship among the dimensionless numbers obtained from this equation.* Analytical methods provide the initial description of phenomena in a form of complicated differential equations that determine interrelations among values in formulating a problem. However, their solution to design relationships is usually not achieved because of the complexity of the problem. Therefore, although a purely analytical investigation remains, which, as a rule, is the most desirable approach to problem solving, it is seldom applied to engineering solutions. On the other hand, purely experimental approaches without knowledge of at least initial theoretical expressions are often doomed to failure. Blind or brute force experimentation often results in a tremendous volume of data, much of which is extraneous and unrelated to the desired solution.

Similarity theory brings to the *experimental* solution of the problem *physical laws* in the form of initial equations describing the process. The transition to generalized variables and modeling significantly facilitates and accelerates this solution.

Any dimensionless expression (despite the empirical method of its derivation in an explicit form) has a definite physical meaning because it reflects the laws of nature expressed by the initial system of physical equations. Modeling based on dimensional analysis produces only approximate solutions because only the most important parameters describing the phenomenon are included in the final expressions. Each dimensionless group in the generalized equation reflects some of the aspects of the process, and the total equation attempts to approximate the behavior of the process as a whole. The technique of treating experimental data in terms of power law expressions is illustrated in the following example.

Example 2.1

A process may be described by a relationship in which $\pi = 2$. Develop a relationship for the function $K_1 = f(K_2)$.

Solution

We may assume a general form of the relationship to be $K_1 = CK_2^n$, where C and n are the unknowns. Taking the logarithm of this expression, we obtain

$$\log K_1 = \log C + n \log K_2 \qquad (2.35)$$

This is the equation of a straight line on log–log coordinates. If we plot the experimental data, and they can be correlated by a straight line, then the slope will determine the coefficient n, and C can be computed for any point on the line. If the data cannot be correlated by a straight line, then a portion of the phenomenon is not accounted for in Equation (2.35), and further clarification of the system's physics is needed.

Kirpichev-Gukhman's Theorem

The first and second theorems neither provide information needed to establish which parameters should be known to satisfy similarity, nor provide a basis for designing the model and the experiments.

This third theorem, along with the first two, serve as similarity indicators and address the issue concerning the field of application of similarity dimensionless equations. It should be obvious that when experimental results are represented in the form of a dimensionless function, its application may only be extended to a group of similar phenomena having common properties.

The dimensionless equation is correct only within the limits of maintaining similarity, which is characterized by definite intervals of changing dimensionless groups K_2, K_3, K_4 in the final expression. Beyond the investigated intervals of changing dimensionless groups, the generalization loses its validity. Hence, an expression derived by the methods outlined in this chapter generally should never be extrapolated much beyond the range in which the important dimensionless groups have been studied experimentally.

This third theorem states that *phenomena are similar if they can be described by the same system of differential equations and have similar conditions of uniqueness.* Consequently, the phenomena are physically similar if they belong to the same class and enter the same group of phenomena, which differ only by a scale of physical values. The fulfillment of conditions of uniqueness in dimensionless treatment is as valid as the determination of uniqueness in analytical solutions of physical differential equations.

Differential equations describe a variety of phenomena of a given class that are based on common physical laws. For example, the class of phenomena of heat propagation is subordinated to the law of heat conduction. However, the general equation does not reflect specific indications of particular phenomena of a given class, for example, heating of a pellet in a catalyst layer. Therefore, many solutions correspond to the initial equation which is *many-valued*. The engineer is generally interested in a specific phenomenon within a given class, and most often under the observed operating conditions in a specific piece of equipment. Therefore, it is necessary *to select from a set* of possible solutions to an initial equation (or system of equations) *one* solution corresponding to the analyzed phenomenon.

Therefore, *additional conditions of uniqueness (boundary conditions)* are added to the problem to obtain a single-valued solution. The boundary conditions include the following:

1. Information on geometric properties of a system (the shae and size of a piece of equipment and the working volume)
2. Data on the physical properties of products and materials composing the system to be investigated (conductivity of heat, heat capacity of equipment walls, viscosity, density of working media, etc.)
3. Data on the condition of a system at its boundary (boundary or space boundary conditions) and on the system's interaction with the surrounding medium (intensity of heat or mass transfer, distribution of temperatures or concentrations on a surface, etc.)
4. Data on the system's condition in the initial and final moments of a process (i.e., time conditions)

The initial system of physical equations, together with boundary conditions, *uniquely* determine a specific phenomenon of a given class. The solution of such a system, in the form of an equation relating to the basic parameters of a process, contains information of practical importance. Unfortunately, the analytical solution of the equation system is not achieved because of its complexity. In this case, one has recourse to the method of generalized variables, i.e., to the theory of similarity.

It is evident that dimensionless numbers should be derived not only from a basic physical equation of a process, but also from equations of conditions of uniqueness. The similarity conditions of uniqueness on the boundary of a system and during a process will determine the physical similarity of processes for the entire system volume. The similarity of conditions of uniqueness will manifest *in coincidence of numerical values* of dimensionless numbers in similar processes.

PRINCIPLES AND METHODS OF MODELING UNIT OPERATIONS

There are two general approaches to modeling, namely, physical and mathematical. In *physical* modeling, the phenomenon occurring in the true or prototype system is approximated under laboratory conditions. For example, in a small model of a heat exchanger, the heat transfer process is reproduced between real working bodies, i.e., heating steam and product. In *mathematical* modeling, or modeling by analogy, the model reproduces a physically different phenomenon, described by the same kind of equations as the phenomenon in the prototype. In this approach, a property called *equation isomorphism* is involved. This property dictates that the same system of equations may be used to describe other phenomena in nature. The property is especially useful in describing fields of temperatures, concentrations, velocities, etc.

Physical and mathematical modeling are based on a unified method of generalized variables. Both methods solve the same problems but accomplish it differently. Each *differs* by specific requirements to initial data and some other properties at a considerable generality of the input information. Thus, for physical modeling it is sufficient to have physical equations in the most general form. In contrast, this is unacceptable for both analytical solutions and those achieved through the use of a mathematical model.

Mathematical modeling is only possible when the equations can be transformed conveniently to a useful solution form. Both methods permit differential equations to be solved; however, it is often easier to alter process parameters and analyze their influence on the performance in a prototype.

Regardless of the modeling approach, one must observe similarity of the following conditions of uniqueness in accordance with the third theorem of similarity:

1 *Geometric similarity* is necessary for physical and mathematical similarity of physical fields. This conditon is eliminated in the mathematical modeling of systems with centered parameters.

2 *Time similarity* must be maintained in both physical and mathematical modeling; the time correspondence is called *homochronicity*. The physical meaning of *homochronicity* is the history of a process changing in time, including transient situations in which the similarity of physical values in a model and in a prototype occurs in *compatible* time moments from the beginning of the process.

3 *Similarity of physical values* should be maintained in all cases of modeling. In *mathematical* modeling, the scale factors have dimension because the model presentation of dimensional physical values of the prototype are values with other dimensions.

4 *Similarity of initial conditions* is also mandatory for both physical and mathematical modeling because the process being developed in time is determined by the *properties* of the process itself, as well as by *initial conditions*. In some cases, the system may be stable, while in others it is unstable, which is unacceptable.

5 *Similarity of boundary conditions* is also necessary for "outlining" the sphere in which the phenomenon proceeds. Other phenomena, independent of the main process under investigation, may occur on the geometric boundaries of the zones. These related phenomena could influence the overall process indirectly. For example, heat transfer from the surface of a body to the atmosphere is independent of heat conduction inside the body but has an influence on the rate of cooling. For boundary conditions, which change with time, it is necessary to maintain time similarity of corresponding variables, which enter into the boundary conditions.

The conditions of uniqueness determine *the scale* of variables and physical parameters of the process according to the requirements of the first theorem. Characteristic parameters should be chosen so that the similarity indicator is equal to unity.

The exact observance of conditions of uniqueness is difficult and may be achieved only in single cases. The fulfillment of uniqueness conditions is especially difficult in those cases in which different and interlinked processes proceed (for example, heat and mass transfer). The requirements of *exact* similarity of conditions of uniqueness would mean providing corresponding equality to all dimensionless groups, including those variables that are unknown. In a practical sense, this is not feasible; therefore, *total similarity* of a model and a prototype is never really achieved, and one has recourse to an *approximate* modeling of the most important dimensionless numbers.

Short-cuts to physical modeling have, on occasion, been applied successfully to experimental investigations. The selection of a specific method is determined by an actual problem. Often, engineering solutions to problems are based on experimental studies of only those parameters or perhaps pieces of equipment that are most important. Common methods of approximate physical modeling are outlined below:

1. Experimental studies of an equipment's *element* can sometimes be made, rather than simulating the entire unit. As an example, the investigation of heat transfer in a multitubular commercial exchanger could be simulated in a single pipe or small pipe bundle. With this type of approach, it is possible to model the prototype design in full scale (that is, a tube simply could be a vertical "extraction" of the commercial unit).

2. Analysis of the prototype is possible through local modeling of complex configurations of the full-scale unit. With this approach, different working portions of the prototype could be modeled separately. Data obtained could be summed graphically by constructing the volume field of measured values. The similarity is thus provided during the experiment but not over the entire volume of the apparatus. Instead, modeling is performed locally and over different time intervals.

3. Isothermal modeling of hydrodynamic processes can be performed by *averaging* physical values, depending on the temperature field in a prototype.

4. Modeling may be performed in the so-called *auto-modeling region* when one of the basic dimensionless numbers of the process becomes unimportant ("confluent"), and its change does not influence the measured values. Then the condition of equality of this number in model and prototype may not be fulfilled.

5. The actual working media of the prototype can sometimes be substi-

tuted with a "modeling" media. One example is substituting water for air in flow studies for furnaces. It is possible in some cases not to fulfill some conditions of uniqueness. In a number of chemical processes, there is no need for geometric similarity of a system. For example, it is possible to create a chemical model of the sea in a glass of water by simulating only one dimensionless parameter—concentration. Geometric similarity of model and prototype is not always a sufficient condition of physical similarity. Consider a group of wooden and iron spheres. All the spheres are geometrically similar, but the wooden spheres are not physically similar to the iron ones. It is evident that the more complicated the process, the more phenomena act in the model at the same time and, thus, the more difficult it is to construct an accurate mathematical description.

The confidence in similarity of phenomena is especially important when a prototype is to be scaled up from a small model. Often, a stepwise approach to design is taken; that is, "micromodels" on bench-scale ("models of models"), can be studied first to understand phenomena and to more carefully design larger-scale models. This approach often provides more confidence in scaling up equipment, especially if the conditions of similarity between prototype and model cannot be strictly maintained. This approach also has the advantage of identifying parameters that do not play dominant roles in the phenomenon and, therefore, may be neglected in experimental studies on larger models.

When changing model scales, the operating conditions are often changed. Typically, the smaller the model, the less accurate the prototype simulation. By transferring to larger models, the process becomes more complicated; therefore, the number of factors to be investigated increases.

Modeling usually can be performed in the following order:

1 A mathematical description of the process is composed in terms of the physical equations and conditions of uniqueness.
2 Dimensionless groups are identified along with the specific dimensionless number containing the design parameter of interest. The design dimensionless number is an implicit function of other dimensionless numbers, which are referred to as the *determining dimensionless groups.*
3 From the condition of equality for dimensionless groups between model and prototype scale, factors are selected for each physical value.
4 On the basis of the above information, a model is constructed whose working volume is geometrically similar to the prototype. The scale of the model is determined from considerations of the size and produc-

Table 2.1. Various Dimensionless Groups and Their Physical Significance.

Dimensionless Group	Symbol	Definition	Significance, Ratio of
Reynolds Number	Re	$\dfrac{\varrho v L}{\mu}$ ϱ = fluid density v = fluid velocity μ = fluid viscosity L = characteristic dimension	$\dfrac{\text{Inertial force}}{\text{Viscous force}}$
Froude Number	Fr	$\dfrac{v^2}{Lg}$	$\dfrac{\text{Inertial force}}{\text{Gravitational force}}$
Euler Number	Eu	$\dfrac{p}{\varrho v^2}$ p = pressure	$\dfrac{\text{Pressure}}{2 \times \text{Velocity head}}$
Mach Number	Ma	$\dfrac{v}{v_c}$	$\dfrac{\text{Fluid velocity}}{\text{Velocity of sound}}$
Weber Number	We	$\dfrac{\pi L v^2}{\sigma}$ σ = surface tension	$\dfrac{\text{Inertial force}}{\text{Surface tension force}}$
Drag Coefficient	C_D	$\dfrac{(\varrho - \varrho')Lg}{\varrho v^2}$	$\dfrac{\text{Gravitational force}}{\text{Inertial force}}$

Table 2.1. (continued).

Dimensionless Group	Symbol	Definition	Significance, Ratio of
		ϱ = density of object	
		ϱ' = density of surrounding fluid	
Fanning Friction Factor	f	$\dfrac{D}{L}\dfrac{\Delta P}{2\varrho v^2}$	$\dfrac{\text{Wall shear stress}}{\text{Velocity head}}$
		D = pipe diameter	
		L = pipe length	
Nusselt Number (heat transfer)	Nu	$\dfrac{hL}{k}$	$\dfrac{\text{Total heat transfer}}{\text{Conductive heat transfer}}$
		h = heat transfer coefficient	
		k = thermal conductivity	
Prandtl Number	Pr	$\dfrac{C_p \mu}{k}$	$\dfrac{\text{Momentum diffusivity}}{\text{Thermal diffusivity}}$
		C_p = heat capacity	
Peclet Number (heat transfer)	Pe	$\dfrac{C_p \varrho v L}{k} = RePr$	$\dfrac{\text{Bulk heat transport}}{\text{Conductive heat transfer}}$
Grashof Number	Gr	$\dfrac{gb^3 \varrho^2 \beta \Delta T}{\mu^2}$	$Re \times \dfrac{\text{buoyancy force}}{\text{viscous force}}$

(continued)

Table 2.1. (continued).

Dimensionless Group	Symbol	Definition	Significance, Ratio of
Grashof Number (continued)		β = coefficient of expansion	
		ΔT = temperature difference	
		b = height of surface	
Stanton Number	St	$\dfrac{h}{\varrho v C_p} = NuRe^{-1}Pr^{-1}$	$\dfrac{\text{Heat transferred}}{\text{Thermal capacity of fluid}}$
J Factor for Heat Transfer	j_H	$\dfrac{h}{\varrho v C_p}\left(\dfrac{C_p \mu}{k}\right)^{2/3}$	Proportional to $NuRe^{-1}Pr^{-1/3}$
Nusselt Number (mass transfer)	Nu	$\dfrac{k_c L}{\mathscr{D}}$	$\dfrac{\text{Total mass transfer}}{\text{Diffusive mass transfer}}$
		k_c = mass transfer coefficient	
		$\mathscr{D}$ = molecular diffusivity	
Schmidt Number	Sc	$\dfrac{\mu}{\varrho \mathscr{D}}$	$\dfrac{\text{Momentum diffusivity}}{\text{Molecular diffusivity}}$
Peclet Number (mass transfer)	Pe	$\dfrac{Lv}{\mathscr{D}} = ReSc$	$\dfrac{\text{Bulk mass transport}}{\text{Diffusive mass transport}}$
J Factor for Mass Transfer	j_D	$\dfrac{k_c}{v}\left(\dfrac{\mu}{\varrho \mathscr{D}}\right)^{2/3}$	Proportional to $NuRe^{-1}Sc^{-1/3}$

tivity of the prototype, providing the necessary velocities, rates, temperatures and other values of the working environment.

5 Experiments should be planned such that dimensionless numbers in the model change over the same limits as in the prototype. By fulfilling this last requirement, the characterizing phenomenon studied in the model would be proportional to the prototype.

The method of dimensional analysis identified dimensionless groups; however, their physical significance is not always evident. The use of differential equations allows one to physically interpret the derived dimensionless groups but still does not provide information on the fundamental mechanism of the process. This can be evaluated only from experimental observation. Table 2.1 summarizes the physical significance of common dimensionless groups.

NOTATION

a = acceleration, m/s²
C, c = dimensionless velocity variable
D, d = diameter, m
$\mathscr{D}$ = molecular diffusivity, m²/s
Fr = Froude number
f = force, N
g = gravitational acceleration, m/s²
i = parameter scale factor
$\tilde{j}$ = nondimensional parameter (transformation variable)
K = dimensionless velocity ratio, see Equation (2.21)
L, ℓ = length, m
M, m = mass, kg
N = number of physical values
Ne = Newton number, see Equation (2.10)
n = number of fundamental units
P = pressure, N/m²
Re = Reynolds number
Stk = Stokes number, see Equation (2.18B)
T, t = time, s
v, w = velocity, m/s

Greek Symbols

γ = specific weight, N
θ = time constant, or angle, in radians
Λ = scale factor

μ = viscosity, cp
ν = kinematic viscosity, m²/s
π = number of functional dimensionless groups, refer to Equation (2.30)
ϱ = density, kg/m³

REFERENCES

Braines, Ya. M. "Podobie i Modelirovanie v. Khimicheskoy i Neftekhimicheskoy Technologi," *Gostoptekhizdat, Moscow* (1961).

Buckingham, E. "On Physically Similar Systems: Illustration of the Use of Dimensional Equations," *Phys. Rev.* 4:345 (1914).

Focken, C. M. *Dimensional Methods and Their Application.* London:Arnold Press (1952).

Gukhman, A. A. *Introduction to the Theory of Similarity.* New York:Academic Press, Inc. (1965).

Ipsen, D. C. *Units, Dimensions, and Dimensionless Numbers.* New York:McGraw-Hill Book Co. (1960).

Kirpichev, M. V. Teorya podobya, Moscow, Is-vo, *AN USSR* (1953).

Klinkenberg, A. "Dimensional Systems and Systems of Units in Physics with Special Reference to Chemical Engineering," *Chem. Eng. Sci.*, 4(130):167 (1955).

Langhaar, H. L. *Dimensional Analysis and Theory of Models.* New York:John Wiley & Sons, Inc. (1951).

Lighthill, M. J. In *Laminar Boundary Layer.* L. Rosenhead, ed., Oxford, England:Clarendon Press (1963).

Pankhurst, R. C. *Dimensional Analysis and Scale Factors.* New York:Van Nostrand Reinhold Co. (1964).

Sedov, L. I. *Similarity and Dimensional Methods in Mechanics.* New York:Van Nostrand Reinhold Co. (1959).

Venikov, V. A. and A. V. Ivanov-Smolenski. "Fisicheskoe Modelirovanie," *Gosenergoizdat, Moscow* (1956).

CHAPTER 3

Hydraulic Processes

INTRODUCTION

Most unit operations of chemical engineering involve the motion of fluids. These fluids may comprise liquids, gases, vapors, or combinations of these. The manner in which fluids interact depends on the nature of the specific unit operation, with applications ranging from fluid transport to mixing, as well as separation of nonhomogeneous mixtures by precipitation, filtration, centrifugation, extraction, distillation, absorption, etc.

These processes are determined by the laws of hydromechanics and, hence, are called *hydromechanical processes*. The laws of hydromechanics are studied in hydraulics, which consists of *two parts:* hydrostatics and hydrodynamics.

Hydrostatics is concerned with the equilibrium state of fluids under stationary conditions, whereas *hydrodynamics* involves the laws of fluids in motion. Classification of hydrodynamic processes may be established on the basis of flow patterns. On this basis, three groups of hydrodynamic problems are defined in this volume—*internal, external,* and *mixed.*

Internal hydrodynamic problems are related to fluid motion in pipes, channels, and equipment. External problems of hydrodynamics cover the motion of bodies through fluid media and involve the analysis of fluid motion in relation to simple bodies (e.g., mechanical mixing, particle sedimentation in liquids or gases, etc.). Mixed problems of hydrodynamics involve the analysis of fluid motion caused by complex interaction with solid obstacles, for example, liquid motion through a grain layer of a solid material or liquid flowing inside channels of complicated shapes while simultaneously flowing around solid particles.

It is essential to obtain a thorough understanding of the hydromechanical processes of any specific unit operation under consideration because hydrodynamic characteristics establish the rates of heat and mass transfer and of chemical reactions.

BASIC DEFINITIONS

Liquid and solid states of matter are the *condensed phases* of gases. The term "condensed phases" emphasizes the high concentration of solid and liquid molecules in comparison to the low density of gases. The volume per mole of gas is very large in comparison to liquids. At conditions of standard temperature and pressure (STP), any gas occupies 22,400 cm³/g-mol. In contrast, most liquids occupy between 10 and 100 cm³/g-mol. That is, the molar volume of a liquid is 500–1000 times smaller than in its gaseous state.

From the kinetic theory of gases, the gas/liquid volume ratio is as large as 1000; therefore, the ratio of distances between molecules in gas to molecules in liquid is $\sqrt[3]{1000}$, or 10. That is, the distance between gas molecules is ten times greater than in a liquid. For a liquid, the average distance between molecules is on the order of a molecular diameter. This large difference in molecular distance between gases and liquids accounts for the drastic difference in properties between these two states of matter. Intermolecular forces, called van der Waals forces, decrease dramatically with distances between molecules. For liquids, these forces are on the order of 10^6 times larger than for gases.

These large differences on the molecular scale cause liquids and gases to display distinct properties when subjected to the same environment. For example, the volume of a fluid as a function of temperature (pressure maintained constant) may be described by the following expression:

$$V = V_0(1 + \varkappa T) \qquad (3.1)$$

where V_0 is the volume of the fluid at some temperature, usually 0°C, and $\varkappa$ is the coefficient of thermal expansion. Gases follow a general dependency similar to Equation (3.1); however, $\varkappa$ is approximately constant for most gases. For liquids, $\varkappa$ depends on the specific liquid.

This dependency of liquid volume on pressure may be expressed in terms of the coefficient of compressibility β as follows:

$$V_0 = \hat{V}_0[1 - \beta(P - 1)] \qquad (3.2)$$

$\hat{V}_0$ is the volume of the fluid at STP. The coefficient β is observed to be constant over a wide range of pressures for a particular material, but is different for each substance and for the solid and liquid states of the same material. Equation (3.2), an equation of state for liquids, states that volume decreases linearly with pressure. In its gaseous state, volume is observed to be inversely proportional to pressure. For a liquid, β is typically 10^{-6} atm^{-1}, whereas for gases it is significant. As an example, if water in its liquid state is subjected to a pressure change from 1 to 2 atm, Equation (3.2) predicts less than $10^{-3}\%$ reduction in volume. However, when this same pressure differential is applied to water as vapor, a volume reduction in excess of two occurs. Because of this property of insignificant volume changes over moderate pressure changes, liquids are referred to as *incompressible fluids,* while gases are called *compressible fluids.*

The terms compressible and incompressible are relative, however, because liquids can change appreciably if conditions are changed over wide limits. Gases also may behave as incompressible fluids if they are subjected to very small changes in pressure and temperature.

The term *liquid* is used in this text in a broad sense. Under the term "liquid," it is necessary to understand substances that possess fluidity. Unlike gases, liquids will not completely fill a volume of specified boundaries.

The general laws of motion describe both liquids and gases, provided gas velocities do not exceed the speed of sound. For liquids, the general laws of equilibrium and motion are expressed in terms of differential equations, where liquids are considered to be homogeneous continuous media. To derive the governing theorems of fluid mechanics, it is necessary to introduce the hypothesis of the *ideal* liquid, in contrast to *real* or *viscous* liquids. That is, for our initial discussions, a real fluid is one that is absolutely incompressible and does not undergo changes in density with variations in temperature. In addition, we assume the fluid is not viscous.

PHYSICAL PROPERTIES OF FLUIDS

Density and Specific Gravity

Fluids may be characterized by the following basic physical properties: density (or specific weight), viscosity, and surface tension.

The mass of liquid per unit volume is called *density* and is denoted by the Greek symbol ϱ:

$$\varrho = \frac{m}{V} \left[\frac{\text{kg}}{\text{m}^3} \right] \qquad (3.3\text{A})$$

where

$m =$ liquid mass (kg)
$V =$ liquid volume (m³)

The weight per unit of fluid volume is called specific gravity:

$$\gamma = \frac{G}{V} \left[\frac{\text{N}}{\text{m}^3} \right] \qquad (3.3\text{B})$$

where

G = weight of the fluid (N)

Note that mass m equals G/g, where g is the acceleration due to gravity. Substituting Equation (3.2) into Equation (3.3) and denoting G/V by γ, we obtain

$$\gamma = \varrho g \qquad (3.4)$$

The density of gases is highly dependent on temperature and pressure. The relationship of temperature, pressure, and density for gases is expressed by the *equation of state* for ideal gases:

$$PV = \frac{mRT}{M} \qquad (3.5)$$

where

P = pressure (N/m²)
V = gas volume (m³)
m = mass of gas (kg)
R = universal gas constant, R = 8314/k-mol (K)
T = temperature (K)
M = molecular weight of the gas

Equation (3.5) may be rearranged to solve for pressure:

$$P = \frac{m}{V}\frac{RT}{M} = \frac{\varrho RT}{M} \qquad (3.6)$$

The volume per unit of mass of gas is called *specific volume:*

$$\dot{\nu} = \frac{V}{m}\left[\frac{\text{m}^3}{\text{kg}}\right] \qquad (3.7)$$

Specific volume is the inverse of density, i.e., $\dot{\nu} = 1/\varrho$, and, consequently, Equation (3.5) may be rewritten as follows:

$$P\dot{\nu} = \frac{RT}{M} \qquad (3.8)$$

The following example illustrates the use of these definitions.

Example 3.1

Determine the density of gaseous ammonia at a pressure $P = 26$ atm gauge and a temperature of 16°C.

Solution

The absolute pressure is

$$P = 26 + 1 = 27 \text{ kg/cm}^2 = 265 \times 10^4 \text{ N/m}^2$$

The molecular weight of NH_3 (ammonia) is $M = 17$.
From Equation (3.6) the density of NH_3 is

$$\varrho = \frac{PM}{RT} = \frac{(265 \times 10^4)(17)}{(8314)(273 + 16)} = 1.87 \text{ kg/m}^3$$

Viscosity

When a real fluid is set in motion, forces of internal friction arise acting in opposition to the direction of flow. This property of resisting motion is called *viscosity*.

Consider a liquid flowing through a cylindrical tube. The fluid can be visualized as concentric rings or layers of fluid, as illustrated in Figure 3.1. If a certain layer has a velocity w then the next layer has a velocity $w + \Delta w$. Experimental observations reveal that the velocity of the layers decreases from the axis to the tube wall, where the velocity is equal to zero. For displacement of one layer relative to the other, it is necessary to apply a force proportional to the surface contact area of the layers. The force f per unit of shearing plane F is called the *shear stress* and is denoted by the following:

$$\tau = \frac{f}{F} \qquad (3.9)$$

Newton's law of viscosity states that the shear stress is proportional to the velocity gradient across the tube through a proportionality constant μ:

$$\tau = \mu \frac{dw}{dn} \qquad (3.10)$$

dw/dn is the velocity gradient normal from the tube wall (i.e., it is the relative velocity changing over unity of distance between layers in a direction perpendicular to the liquid flow).

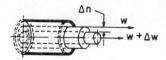

FIGURE 3.1. Conceptual interpretation of fluid flowing full through a tube.

The proportionality constant in Equation (3.10) depends on the physical properties of the fluid and is called the *coefficient dynamic viscosity* or, simply, *viscosity.*

Substituting for τ from Equation (3.9), we obtain the dimensions of viscosity:

$$[\mu] = \left[\frac{f}{F\dfrac{dw}{dn}}\right] = \left[\frac{N}{m^2 \dfrac{m/s}{m}}\right] = \left[\frac{N \times s}{m^2}\right]$$

Note that $(N) = (kg\text{-}m/s^2)$, whence we obtain another form of viscosity:

$$[\mu] = \left[\frac{\dfrac{kg\text{-}m}{s^2} \times s}{m^2}\right] = \left[\frac{kg}{m \times s}\right]$$

In the cgs system, the unit of viscosity is the poise (ps). A poise refers to a force of 1 dyne that displaces liquid layers having a surface area of 1 cm², each situated at 1 cm from each other with a relative velocity of 1 cm/s. That is,

$$(ps) = \left[\frac{dyne \times s}{cm^2}\right] = \left[\frac{g \times cm}{s^2} \times \frac{s}{cm^2}\right] = \left[\frac{g}{cm \times s}\right]$$

The unit of viscosity equal to 0.01 ps is called a *centipoise* (cps). Typical values for different fluids as a function of temperature and pressure may be found in standard handbooks.

The ratio of viscosity μ to density ϱ is the *coefficient of kinematic viscosity* or, simply, *kinematic viscosity:*

$$\nu = \frac{\mu}{\varrho} \qquad (3.11)$$

The dimensions of kinematic viscosity are

$$[\nu] = \left[\frac{N \times s/m^2}{kg/m^3}\right] = \left[\frac{\frac{kg \times m}{s^2} \times \frac{s}{m^2}}{\frac{kg}{m^3}}\right] = \left[\frac{m^2}{s}\right]$$

The following two examples illustrate the calculation of viscosity for a gas and a liquid and employ principles of physical chemistry. The reader who wishes to review these principles in more depth should consult the work of Cheremisinoff (1981) or Castellan (1971).

Example 3.2

Calculate the viscosity of sulfur dioxide gas at a temperature of 300°C and atmospheric pressure. The viscosity of SO_2 at 20°C and 150°C is equal to 1.26×10^{-2} and 1.86×10^{-2} cp, respectively, and its critical temperature and pressure are as follows: $T_{cr} = 430K$ at $P_{cr} = 77.7$ atm. Compare the calculated value with an experimentally determined value of 2.46×10^{-2} cp.

Solution

1. To calculate viscosity, the following equation is used:

$$\mu = 6.3 \times 10^{-4} \frac{M^{1/2} P_{cr}^{2/3}}{T_{cr}^{1/6}} \cdot \frac{T_{red}^{3/2}}{T_{red} + 0.8} \tag{3.12}$$

where

$$T_{red} = T/T_{cr}$$

The subscript "red" refers to a *reduced* state:

$$M_{SO_2} = 64; \quad T_{red} = \frac{(300 + 273)}{430} = 1.335$$

Substituting these values into the above equation, we obtain the value of viscosity:

$$\mu = (6.3 \times 10^{-4}) \frac{(64)^{1/2} \times (77.7)^{2/3}}{(430)^{1/6}} \times \frac{(1.335)^{3/2}}{1.335 + 0.8} = 2.41 \times 10^{-2} \text{ cp}$$

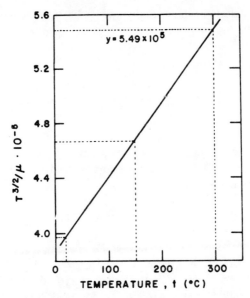

FIGURE 3.2. A plot of the function $y = T^{3/2}/\mu$ versus temperature, t. The plot is used in evaluating the viscosity of SO_2 gas by method 2 in Example 3.2.

2. We now construct a graph of $y = T^{3/2}/\mu$ versus t using two known values of μ:

$$\text{at } t_1 = 20°C; \quad y_1 = \frac{T^{3/2}}{\mu} = \frac{(239)^{3/2}}{1.26 \times 10^{-2}} = 3.98 \times 10^5$$

$$\text{at } t_2 = 150°C; \quad y_2 = \frac{T^{3/2}}{\mu} = \frac{(423)^{3/2}}{1.86 \times 10^{-2}} = 4.68 \times 10^5$$

The plot is shown in Figure 3.2. From the plot in Figure 3.2, $y = 5.485 \times 10^5$ at 300°C. Therefore:

$$\mu = \frac{T^{3/2}}{y} = \frac{(573)^{3/2}}{5.485 \times 10^5} = 2.49 \times 10^{-2} \text{ cp}$$

The same value of y may be obtained by using the equation of a straight line passing the points t_1, y_1 and t_2, y_2.

3. We now determine the viscosity at $t = 300°C$, assuming we know μ_1 at $t = 150°C$. In this case, we use the following equation:

$$\mu = \mu_1 \left(\frac{T_{red}}{T_{red_1}}\right)^{3/2} \times \frac{T_{red_1} + 0.8}{T_{red} + 0.8}$$

(3.13)

$$\mu = 1.86 \times 10^{-2} \left(\frac{573}{423}\right)^{3/2} \times \frac{423/430 + 0.8}{573/430 + 0.8} = 2.44 \times 10^{-2} \text{ cp}$$

Note that the errors associated with computing the viscosity of SO_2 by the different methods outlined above are small.

Example 3.3

Calculate the viscosity of acetic acid at $t = 40°C$. The density of the acid is 1.027 g/cm³ at 40°C. The viscosities of the acid at $t_1 = 20°C$ and $t_2 = 100°C$ are $\mu_1 = 1.22$ cp and $\mu_2 = 0.46$ cp, respectively. Compare the computed values of μ with the experimental value of 0.9 cp.

Solution

1. Use the following equation:

$$\log(\log 10\mu) = K \frac{\varrho}{M} - 2.9$$

(3.14)

where

μ = liquid viscosity (cp)
ϱ = liquid density (g/cm³)
M = molecular weight
K = constant determined from the following equation:

$$K = \Sigma m I_a + \Sigma I_c$$

(3.15)

where

mI_a = product of the number of atoms m of a given element in a molecule times its corresponding atomic constant I_a
I_c = constant characterizing the molecular structure; it is determined by the atoms' grouping and bonds among them

Table 3.1. Values of Constant I_c.

Compound, Grouping or Bond	Constant I_c	Compound, Grouping or Bond	Constant I_c
Double Bond	−15.5		
Five Ring C Atoms	−24.0	$CH_3-\underset{\underset{O}{\|}}{C}-R$	+5.0
Six Ring C Atoms	−21.0		
Substitution in Six-Ring Member In ortho- and para-position	+3.0	$-CH-CHCH_2X$ [a]	+4.0
In meta-position	+1.0		
$\underset{R}{\overset{R}{>}}CHCH\underset{R}{\overset{R}{<}}$	+8.0	$\underset{R}{\overset{R}{>}}CHX$ [a]	+6.0
$R-\underset{\underset{R}{\|}}{\overset{\overset{R}{\|}}{C}}-R$	+13.0	−OH −COO− −COOH −NO$_2$	+24.7 −19.6 −7.9 −6.4
$R-\underset{\underset{O}{\|}}{C}-H$	+10.0		

[a]Electronegative group.

Values of constant I_a for different elements are as follows:

Element	H	C	O	N	Cl	Br	I
Constant I_a	2.7	50.2	29.7	37	60	79	110

Typical values of I_c are given in Table 3.1.

The influence of temperature on liquid viscosity may be determined from viscosity values at two different temperatures by using a comparison with a reference liquid. This can be stated as follows:

$$\frac{t_1 - t_2}{\theta_1 - \theta_2} = K \frac{t_1 - t}{\theta_1 - \theta} \quad (3.16)$$

where

t_1, t_2 = temperatures of the liquid
θ_1, θ_2 = temperatures of the reference liquid at which its viscosity is equal to the viscosity of the liquid to be compared at t_1 and t_2

The molecular weight of acetic acid is 60.06. The constant K for acetic acid is

$$K = 2I_a(C) + 4I_a(H) + 2I_a(O) + I_c(COOH)$$

$$= 2 \times 50.2 + 4 \times 2.7 + 2 \times 29.7 + 7.9 = 162.7$$

Consequently,

$$\log(\log 10\mu) = 162.7 \frac{1.027}{60.06} - 2.9 = -0.118$$

Hence,

$$\log(10\mu) = 0.762; \; \mu = \frac{5.78}{10} = 0.578 \text{ cp}$$

2. Applying Equation (3.16) and using water as the reference liquid, where $\mu'_{H_2O} = 1.22$ cp and $\mu''_{H_2O} = 0.46$ cp at $\theta_1 = 12.5°C$ and $\theta_2 = 60°C$, respectively, we obtain the following:

$$\theta = \theta_1 + (t - t_1) \frac{\theta_2 - \theta_1}{t_2 - t_1}$$

$$= 12.5 + (40 - 20) \left[\frac{60.7 - 12.5}{100 - 20} \right] = 24.55°C$$

At $\theta = 24.55°C$, $\mu_{H_2O} = 0.909$ cp. Therefore, the viscosity of acetic acid at 40°C is computed to be the same.

PRESSURE AND SURFACE TENSION

Liquid Pressure

A basic property of a static fluid is the pressure it exerts on the walls of its container. Pressure can be interpreted as a surface force exerted by a fluid against the walls and bottom of its container, as well as on a surface of a body submerged in the fluid medium. Liquid pressure exerted against a unit surface area is called *hydrostatic pressure,* or *pressure.* This is stated simply as follows:

$$P = \frac{p}{F} \tag{3.17}$$

where p is the pressure force applied against surface area F.

If a liquid is poured into a vessel, then the pressure force exerted against the vessel floor is equal to the weight of liquid in the vessel:

$$p = FH\varrho g \qquad (3.18)$$

where

F = surface of the vessel's bottom
H = height of liquid column
ϱ = liquid density
g = acceleration due to gravity

From Equation (3.18), the pressure exerted against the bottom of the vessel is

$$P = \frac{FH\varrho g}{F} = H\varrho g \qquad (3.19)$$

Equation (3.19) states that the liquid pressure exerted against the bottom of the vessel is equal to the weight of the column for a bottom area of unity (i.e., $F = 1$). If a pressure P_o is exerted over the upper liquid's exposed surface, then the hydrostatic pressure is

$$P = P_o + H\varrho g \qquad (3.20)$$

The pressure exerted against the vertical or sloped walls of a vessel is not constant over the vessel height. In fact, it should be considered as a limit of the ratio of pressure force Δp to an elemental area ΔF which is subjected to the force action at ΔF approaching zero:

$$P = \lim_{\Delta F \to 0} \left[\frac{\Delta p}{\Delta P} \right] \qquad (3.21)$$

This pressure is normally directed to the area of the walls. If this were not the case, the fluid would no longer be static.

Surface Tension

In a number of unit operations, liquids are contacted with a gas or other immiscible liquids. The surface area of contact between such fluids approaches a minimum value due to the action of surface forces. For example, droplets suspended in a gas (vapor) or gas bubbles dispersed in a liquid medium approach a spherical geometry. This is explained by the fact that molecules at the fluid–fluid interface or close to it are subjected to attraction forces from molecules located inside the bulk fluid medium. Hence, there arises a pressure from within the bulk liquid that is perpendicular to the gas–liquid interface. The action of these forces is manifested by the liquid's attempt to decrease its surface area. In developing a new surface it is necessary to consume some energy.

The work required for the formation of a unit of new surface is called *surface tension* and is denoted by the symbol σ. This work is measured in joules and is related to 1 m² of surface area.

The dimension of surface tension is

$$\sigma = \left[\frac{J}{m^2}\right] = \left[\frac{N \times m}{m^2}\right] = \left[\frac{N}{m}\right]$$

Surface tension may also be considered to be a force acting on a unit length of the interface.

Further explanation of the material presented in this chapter is given by Bennett and Meyers (1964), Bird et al. (1960), Cheremisinoff (1981), Hougen et al. (1954), Perry and Chilton (1973), and Streeter (1971).

NOTATION

F = area, m²
f = force, N
G = weight of fluid, N
g = gravitational acceleration, m/s²
H = height of fluid column, m
I_a, I_c = atomic constants for liquid equation of state
K = constant defined by Equation (3.15)
M = molecular weight
m = mass, kg, or number of atoms
n = normal distance, m
P = pressure, N/m²
P_{cr} = critical pressure, N/m²
P_0 = pressure exerted at liquid surface, N/m²
R = universal gas law constant, J/kg-mol-K
STP = standard temperature and pressure
T = absolute temperature, K or R
T_{red} = reduced temperature, dimensionless
t = temperature, °C or °F
V = volume, m³
V_0 = volume at reference temperature $t = 0°C$, m³
w = fluid velocity, m/s
y = function defined as $T^{3/2}/\mu$, K$^{2/3}$/poise

Greek Symbols

β = coefficient of compressibility, atm^{-1}

γ = specific gravity (strictly defined as weight per unit volume, N/m^3) = ratio of mass of liquid to mass of equal volume of water at STP (for liquids) and = ratio of mass of gas to mass of equal volume of air at STP (for gases)

θ = reference temperature, °C or °F

$\varkappa$ = coefficient of thermal expansion, m/m-K

μ = viscosity, poise = N-s/m^2

ν = kinematic viscosity, m^2/kg

$\dot{\nu}$ = specific volume, m^3/kg

ϱ = density, kg/m^3

σ = surface tension, N/m

τ = shear stress, N/m^2

REFERENCES

BENNETT, C. O. and J. E. Meyers. *Momentum, Heat and Mass Transfer.* 2nd ed., New York: McGraw-Hill Book Co. (1964).

BIRD, R. B., W. E. Stewart and E. N. Lightfoot. *Transport Phenomena.* New York:John Wiley & Sons, Inc. (1960).

CASTELLAN, G. W. *Physical Chemistry.* 2nd ed., Reading, MA:Addison-Wesley Publishing Co. (1971).

CHEREMISINOFF, N. P. *Fluid Flow: Pumps, Pipes and Channels.* Ann Arbor, MI:Ann Arbor Science Publishers, Inc. (1981).

HOUGEN, O. A., K. M. Watson and R. A. Ragutz. *Chemical Process Principles.* Part I, 2nd. ed., New York:McGraw-Hill Book Co. (1954).

PERRY, R. H. and C. H. Chilton, eds. *Chemical Engineer's Handbook.* 5th ed., New York: McGraw-Hill Book Co. (1973).

STREETER, V. L. *Fluid Mechanics.* 5th ed., New York:McGraw-Hill Book Co. (1971).

WEAST, R. C., ed. *Handbook of Chemistry & Physics.* 49th ed., Cleveland, OH:The Chemical Rubber Co. (1968).

CHAPTER 4
Hydrostatics

INTRODUCTION

Hydrostatics is concerned with the conditions of equilibrium when the fluid is in a state of rest. That is, hydrostatics considers the *relative* state of rest of fluids; even if the fluid particles are in motion, they do not undergo displacement relative to each other. For example, a fluid contained in a vessel that is being transported via a truck is said to be in a state of relative rest with respect to any point of reference within the fluid volume. Another example is liquid inside a centrifuge rotating at constant angular speed. When fluids are totally at rest or stationary with respect to the surroundings, no internal frictional forces exist, and the fluid is considered ideal.

At rest, a fluid's shape and volume do not change and, as in the case of a solid object, it displaces the surrounding environment by its own weight. A stationary fluid is subjected to gravity and pressure forces. For a fluid in a relative state of rest, inertia forces of transient motion are also at play. The relationship among the various forces that act against the fluid is determined by equilibrium conditions that can be expressed mathematically by Euler's differential equations.

EULER'S DIFFERENTIAL EQUATIONS

Consider a mass of static fluid in the form of an elementary parallelepiped of volume dV with edges dx, dy, dz, situated parallel to coordinate axes x, y, and z. The system is shown in Figure 4.1. The gravity force acting on this volume is the product of its mass dm times acceleration due to gravity g, i.e., gdm. The force of hydrostatic pressure acting against any side of the parallelepiped is equivalent to the product of pressure p over the side's area. To develop an expression for pressure, we must assume p to be a function of all three coordinates, $P = f(x,y,z)$.

According to basic principles of statics, at equilibrium the sum of the projections onto the coordinate axis of all forces acting on the elementary volume must be zero. If this were not the case, the fluid would be displaced. First we

85

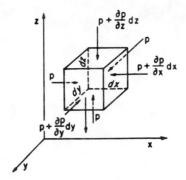

FIGURE 4.1. The system under consideration for developing Euler's equations.

will consider forces acting in the z-direction. The force of gravity acting in the downward direction is parallel to the z-axis.

$$-gdm = -g\varrho dV = \varrho g dxdydz$$

The pressure force acting normal to the lower side of the parallelepiped of the z-axis is $pdxdy$. If the pressure change at a given point in the z-direction is $\partial P/\partial z$, then along edge dz it is $(\partial P/\partial z)dz$.

The pressure acting on the opposite (upper) side is $[P + (\partial P/\partial z)dz]$, and the projection of this pressure force onto the z-axis is

$$-\left(P + \frac{\partial P}{\partial z}dz\right)dxdy$$

Hence, the resulting pressure force on the z-axis is

$$Pdxdy - \left(p + \frac{\partial P}{\partial z}dz\right)dxdy = -\frac{\partial P}{\partial z}dzdxdy$$

and the sum of the forces projected onto the z-axis is equal to zero. That is,

$$-\varrho g dxdydz - \frac{\partial P}{\partial z}dxdydz = 0 \qquad (4.1)$$

and because $dxdydz = dV \neq 0$, we obtain

$$-\varrho g - \frac{\partial P}{\partial z} = 0$$

The projection of gravity forces on the x-axis equals zero. Thus, we may write

$$\varrho dydz - \left(P + \frac{\partial P}{\partial x} dz\right) dydz = 0$$

After simplification, we obtain

$$-\frac{\partial P}{\partial x} dxdydz = 0 \qquad (4.2)$$

or

$$-\frac{\partial P}{\partial x} = 0$$

And, correspondingly, for the y-axis,

$$-\frac{\partial P}{\partial y} dxdydz = 0 \qquad (4.3)$$

or

$$-\frac{\partial P}{\partial y} = 0$$

Thus, the equilibrium conditions of the elementary parallelepiped may be expressed by the following system of equations:

$$\left.\begin{array}{r} -\dfrac{\partial P}{\partial x} = 0 \\[1ex] -\dfrac{\partial P}{\partial y} = 0 \\[1ex] -\varrho g - \dfrac{\partial P}{\partial z} = 0 \end{array}\right\} \qquad (4.4)$$

These expressions are known as Euler's differential equations.

THE BASIC EQUATION OF HYDROSTATICS

Equation (4.4) shows that the pressure of a static fluid varies only vertically. In other words, the pressure at any point on a horizontal plane of the fluid parallelepiped is the same. For the system of equations in Equations (4.4), the partial derivatives $\partial P/\partial x$ and $\partial P/\partial y$ are equal to zero. Consequently, the partial derivative $\partial P/\partial z$ may be substituted by dP/dz:

$$-\varrho g - \frac{dP}{dz} = 0$$

Hence,

$$-dP - \varrho g\, dz = 0$$

or

$$dz + d\left(\frac{P}{\varrho g}\right) = 0$$

For a homogeneous, incompressible fluid, density is constant. Therefore,

$$d\left(z + \frac{P}{\varrho g}\right) = 0$$

After integration, we obtain

$$z + \frac{P}{\varrho g} = \text{const.} \tag{4.5}$$

For any two horizontal planes 1 and 2, Equation (4.5) may be written in the following form:

$$z_1 + \frac{P_1}{\varrho g} = z_2 + \frac{P_2}{\varrho g} \tag{4.6}$$

Equation (4.6) expresses the condition of hydrostatic equilibrium.

For illustration, consider the system shown in Figure 4.2. Let us direct our attention to two liquid particles, one of which is point "A" located at a height

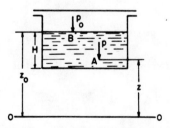

FIGURE 4.2. An example system for developing the basic hydrostatic equation.

z above some arbitrary reference plane 0–0, and the other point "B" on the liquid surface at height z_0. P and P_0 are the pressures at points "A" and "B," respectively. From Equation (4.6) we may write the following:

$$z + \frac{P}{\varrho g} = z_0 + \frac{P_0}{\varrho g} \qquad (4.7)$$

or

$$\frac{P - P_0}{\varrho g} = z_0 - z \qquad (4.8)$$

Height z in Equation (4.6) is commonly referred to as the *leveling height* and has the following units:

$$\frac{P}{\varrho g} = \frac{P}{\gamma} = \left[\frac{\text{N-m}^3}{\text{m}^2\text{-N}}\right] = [\text{m}]$$

The value $P/\varrho g$ is referred to as the static or pressure head. It follows from Equation (4.6) that for any point within the static liquid the sum of the leveling height and pressure head is constant.

The terms of the basic hydrostatic equation have relevance to the energy of the fluid. The term $P/\varrho g$ may be expressed in units of $[(\text{N-m})/\text{N})] = [\text{J/N}]$, which is the *specific energy* of the fluid, i.e., the energy per unit weight [J/N or $(\text{kg}_f\text{-m})/\text{kg}_f$]. The same definition may be used to describe the leveling height if we multiply and divide z by unity of weight.

Thus, leveling height z, also referred to as the *geometric head*, characterizes the *specific potential energy of position* of a given point over a specified, but arbitrary, reference plane. The static head represents the *specific potential energy of pressure* at that point.

The sum of these two energies is equal to the total potential energy per unit weight of fluid. Therefore, the basic equation of hydrostatics is a specific case of the law of conservation of energy, which states simply that the specific potential energy at all points in a static fluid is constant. Equation (4.8) may be rewritten in the following form:

$$P + \varrho g z = P_0 + \varrho g z_0 \qquad (4.9)$$

or

$$P = P_0 + \varrho g (z - z_0) \qquad (4.10)$$

Equation (4.10) is known as Pascal's law, which states that the pressure at any

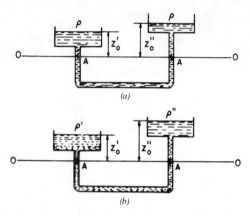

FIGURE 4.3. Equilibrium conditions established in communicating vessels: (a) represents the case of a homogeneous liquid system; (b) represents the case of a heterogeneous (immiscible) liquid system.

point in a static incompressible fluid is transmitted equally to all points within its volume. So, for example, if pressure P_0 at point z_0 is changed by some value, the pressure at any other point in the fluid will also change by an equivalent amount.

APPLICATION OF HYDROSTATIC PRINCIPLES TO MANOMETRIC TECHNIQUES

Efficient process operations rely heavily on accurate measurement and control of the amount of materials entering and leaving equipment. Many measuring and control devices indirectly monitor and regulate flow quantities through the equipment's pressure or some pressure difference. Through Pascal's equation, pressure or a pressure differential can be related to the amount of material present at any one time in a reactor or vessel. This is illustrated best by the *principle of communicating vessels.* Consider two vessels open to the atmosphere and connected to each other as shown by Figure 4.3(a). Both vessels contain the same liquid of density ϱ. Note that the vessel on the right in Figure 4.3(a) is higher than the one on the left. The reference plane 0–0 passes through point A, which represents a position inside the liquid below each vessel. For point A below the left vessel, we have

$$P = P_{atm} + \varrho g z_0'$$

For point A below the right vessel,

$$P = P_{atm} + \varrho g z_0''$$

And as plane 0–0 passes through point A, we recognize that $z_0' = z_0'' = 0$:

$$\left. \begin{array}{c} P_{atm} + \varrho g z_0' = P_{atm} + \varrho g z_0'' \\ \\ z_0' = z_0'' \end{array} \right\} \quad (4.11)$$

At equilibrium, the pressure is equivalent at any point in the liquid. Thus, in communicating vessels (either open or closed) under the same pressure and containing the same fluid, the liquid levels come to rest at the same height. This principle is used, for example, in measuring liquid levels in tanks by the use of water gauges.

Suppose we now drain the system and fill the left vessel with a liquid of density ϱ' and the right with a liquid of density ϱ'', as illustrated in Figure 4.3(b). The liquids are immiscible and, hence, we have a heterogeneous system. Using the same approach, we obtain the following expressions:

$$\varrho' z_0' = \varrho'' z_0'' \quad (4.12)$$

or

$$\frac{z_0'}{z_0''} = \frac{\varrho''}{\varrho'} \quad (4.13)$$

Equation (4.13) states that the levels of the fluids are inversely proportional to their densities for the case of communicating vessels containing a heterogeneous fluid system.

A third situation worth considering is that in which both vessels are filled again with the same liquid of density ϱ, but in which the pressures applied over each liquid surface are different. If the pressure is P' over the left vessel and P'' over the right vessel, then we may write the following:

$$P' + \varrho g z_0' = P'' + \varrho g z_0'' \quad (4.14)$$

or

$$z_0'' - z_0' = \frac{P' - P''}{\varrho g}$$

Equation (4.14) may be used to evaluate how pressure varies with a change in depth or elevation.

Manometers are the simplest devices for measuring pressure and are based on the principle of communicating vessels. The basic manometer consists of a glass tube bent in the shape of the letter "U" and partially filled with a stan-

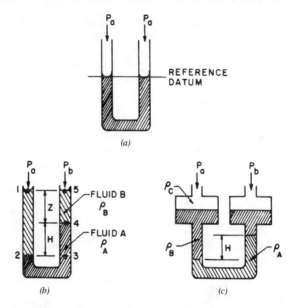

FIGURE 4.4. The basic U-tube manometer used in measuring pressure differences: (a) single fluid manometer open to the atmosphere; (b) a fluid manometer measuring the pressure differential; (c) the size of the manometer legs does not matter in measuring pressure differential; this particular design is a two-fluid manometer.

dard gauge fluid such as water, mercury, or a colored light oil. The basic U-tube manometer is illustrated in Figure 4.4(a). When both ends of the U-tube are open to the atmosphere, the pressure on each side is equivalent; thus, the column of liquid in each leg is balanced exactly. Hence, the liquid surfaces reach equilibrium at the same level.

Let us now consider a case in which the manometer containing a gauge fluid of density ϱ_A(kg/m³) is hooked up to a system containing a fluid of density ϱ_B, in which the system fluid applies a greater pressure to one of the legs. The system is illustrated in Figure 4.4(b). We wish to develop an expression that relates the pressures at the top of each leg, i.e., between P_a and P_b. Starting with the left leg, the pressure at point 2, i.e., at the A–B interface is the following:

$$P_2 = P_a + (z + H)\varrho_B g, \text{ N/m}^2 \qquad (4.15)$$

where H is the displacement of the heavier fluid in m. The pressure at point 3 must be equal to that at point 2 according to our previous discussion. The pressure at point 3 also is equal to the following:

$$P_3 = P_b + z\varrho_B g + H\varrho_A g \qquad (4.16)$$

By combining Equations (4.16) and (4.15) (because $P_3 = P_2$) and rearranging terms, the following expression is obtained:

$$P_a - P_b = H(\varrho_A - \varrho_B)g \qquad (4.17)$$

Two observations should be made of Equation (4.17): (1) the elevation term z cancels out when deriving this expression, and hence, does not require measurement when reading a manometer; and (2) the expression is in SI units [to use Equation (4.17) in English units, divide the RHS by g_c].

Liquid manometers are widely used in both industrial and laboratory applications. they can be employed both as basic pressure measurement devices and as standards of calibration of other instruments.

Only the height of the fluid from the surface of one leg to the surface in the other is the actual height of the fluid opposing and balancing the applied pressure. This is so regardless of the geometry or size of the legs, as illustrated in Figure 4.4(c). Even if the manometer tubes are unsymmetrical, the only occurrence will be that more or less fluid will move from one leg to the other. The liquid height required to achieve equilibrium will depend only on the density of the manometer fluid and its vertical height. Table 4.1 gives equivalent

Table 4.1. Conversion Values for Common Manometer Fluids at 22°C.

1 in water =	0.0360 lb/in^2
	0.5760 oz/in^2
	0.0737 in mercury
	1.2131 in red oil
1 ft water =	0.4320 lb/in^2
	6.9120 oz/in^2
	0.8844 in mercury
	62.208 lb/ft^2
	14.5572 in red oil
1 in mercury =	0.4892 lb/in^2
	7.8272 oz/in^2
	13.5712 in water
	1.1309 ft water
	16.4636 in red oil
1 oz/in^2 =	0.1277 in mercury
	1.7336 in water
	2.1034 in red oil
1 lb/in^2 =	2.0441 in mercury
	27.7417 in water
	2.3118 ft water
	33.6542 in red oil

values of various manometric fluids and demonstrates the versatility of the manometer. For example, when water is used as an indicating fluid, a 10-in. fluid height measures 0.360 psi, whereas the same measured height utilizing mercury corresponds to 4.892 psi. The ratio of these two pressures is 13.57:1, which is the ratio of the specific gravities of the two fluids. Refer back to Equation (4.13). Three types of pressure measurements can be made with manometers:

1 Positive, or gauge, which are pressures greater than atmospheric
2 Negative pressures, or vacuums, which are pressures less than atmospheric
3 Differential pressure, which is the difference between two pressures

Connecting one leg of the U-tube to a source of gauge pressure causes the fluid in the connected leg to depress while the fluid in the vented leg rises. However, if the connection is made to a vacuum, the effect would be to reverse the fluid movement, causing it to rise in the connected leg and recede in the open leg. In obtaining differential pressures, both legs of the manometer are connected so that a pressure change between two points can be measured, as described in the derivation of Equation (4.17). The higher pressure depresses the fluid in one leg, while the lower pressure allows the fluid to rise in the other. The true differential is measured by the difference in height of the fluid in the two legs.

Manometers are available in configurations other than the U-tube to provide greater convenience and to meet different service requirements. Figure 4.5 shows three other types of manometers. Figure 4.5(a) illustrates a well-type manometer. As shown, if one leg of the manometer is increased in area in comparison to the other, the volume of fluid displaced represents little change in the vertical height of the large-area leg compared to the change of height in the smaller-area leg. Thus, it becomes necessary to read only the scale adjacent to the single tube rather than two, as with the U-type.

This well-type design lends itself to the use of direct reading scales that are graduated in appropriate units for the process variable involved. The higher pressure source to be measured must remain connected to the well leg, while the lower pressure source must be connected to the top of the tube. In any measurement, therefore, the source of the higher pressure must be connected in such a manner to cause the manometer fluid to rise in the indicating tube.

The true pressure follows the same principles discussed previously, and the measurement obtained reflects the difference between the fluid surfaces. Obviously, there must be some decrease in the well level; however, this can be compensated for by spacing the scale graduations in the proper amount needed to reflect and correct for the well level drop.

A variety of applications require accurate measurement of low pressures, such as drafts and low differentials in air and gas installations. A common

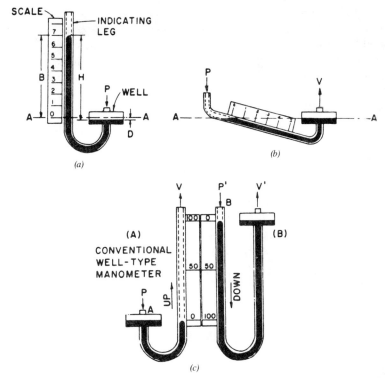

FIGURE 4.5. Alternative configurations of manometers: (a) a well-type manometer; (b) inclined-tube, well-type manometer for low-pressure measurements; (c) dual-tube manometer system for measuring high pressures.

arrangement for these cases is the use of an inclined-tube well manometer as shown in Figure 4.5(c). This design has the advantage of providing an expanded scale. For example, 12 cm of scale length can represent 1 cm of vertical length. Using scale divisions of 0.01 inches of liquid height, an equivalent pressure of 0.000360 psi per division can be detected with water as the manometer fluid.

To measure relatively high pressures, a longer indicating fluid tube is required. Rather than utilizing excessively long manometer tubes for high-pressure readings, a dual-tube manometer arrangement can be used, as shown in Figure 4.5(c). This arrangement provides for reading the full range of the instrument in only one-half the total vertical viewing distance. The system shown in Figure 4.5(c) consists of two separate manometers mounted on a single housing. Manometer (a) is a conventional well design having a zero scale at the bottom and graduated upward. Manometer (b) has a well and zero scale raised to the 100-inch level, with the scale graduated downward. Connecting a positive-pressure source at connections A and B causes the indicating fluid level to rise in tube A and fall in B. Note, however, that these are essentially two separate manometers measuring the same pressure, with one indicating column rising and one falling. The fluid columns pass each other at the 50-inch mark and, as such, it becomes convenient to read the left scale on pressures from 0 to 50 inches progressing upscale and the scale on manometer (b) from 50 to 100 inches downscale. Only the lower 50 inches is required to read the entire range of 100 inches. Gauge pressure connections must be made at connections A and B, while vacuum connections must be made at V and V'.

Another system frequently used is a sealed tube or absolute manometer. The term "absolute pressure" originates from the fact that in a perfect vacuum the complete absence of any gas is referred to as "absolute" zero. For an absolute pressure manometer, the pressure measured is compared to the vacuum or absolute pressure in a sealed tube above the indicating fluid column. The most common type of sealed tube manometer is the conventional mercury barometer, which is used to measure atmospheric pressure. The barometer is a mercury-filled tube more than 76 cm high and immersed in a reservoir of mercury, which is exposed to the atmosphere. The mercury column is supported by the atmospheric or barometric pressure. A variety of processes, tests, and calibrations are based on pressures near or below atmospheric conditions, and these are measured most conveniently with a sealed-tube manometer.

The principle of communicating vessels may also be applied to determining the height of hydraulic seals in different equipment. Figure 4.6 shows such an example in which an emulsion made up of two liquids of different densities continuously enters into vessel 1 through central tube 2, where it separates. The lighter liquid of density ϱ' is skimmed off the fluid surface through nozzle 3, while the heavier fluid of density ϱ'' is drained from below through the U-hydraulic seal 4.

The reference plane "0–0" is specified at the liquid–liquid interface. From Equation (4.13), the height of the hydraulic seal required will be

$$z_0'' = z_0' \frac{\varrho'}{\varrho''}$$

We can assume that the pressures above the liquid surface inside the vessel and at the outlet of the seal are the same.

Figure 4.7 illustrates still another application of manometers. In this example, a manometer (number 3) is used to measure the liquid level in a tank as follows. Pressurized air is passed through tube 2 to the bottom of the tank. The air pressure is increased gradually until it overcomes the resistance of the liquid column in the tank. At the point at which the static head is overcome,

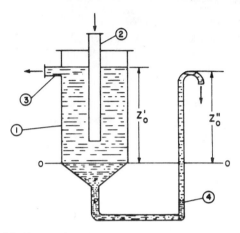

FIGURE 4.6. The use of a hydraulic seal in a continuous liquid separator.

bubbling starts through the liquid and pressure P is maintained constant. From Equation (4.10), the value of the pressure required to overcome the liquid column resistance is

$$P = P_0 + \rho g z_0$$

whence we can solve for the liquid level in the tank:

$$z_0 = \frac{P - P_0}{\rho g} \qquad (4.18)$$

where P_0 is the pressure at the liquid surface. From z_0 and the cross-sectional area of the vessel, one can compute the liquid inventory.

In addition to manometers, the mechanical Bourdon-type pressure gauge is employed extensively as a pressure-measuring device. Its principle of mea-

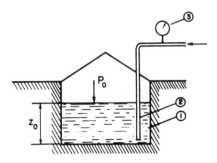

FIGURE 4.7. Application of pneumatic measurement of liquid level in a vessel.

surement is based on a coiled hollow tube housed by the gauge, which tends to straighten out when subjected to an internal pressure. The extent to which this straightening occurs is proportional to the pressure difference between the inside and outside of the gauge. The tube is connected to a point on a calibrated dial.

Further applications of manometers and various pressure-measuring devices used in industry and the laboratory are given by Cheremisinoff (1979, 1981). The following example problems illustrate the use of the principal equations derived in this subsection.

Example 4.1

A simple U-tube manometer is being used to measure the pressure drop of crude oil flowing through an orifice. As will be discussed later, this pressure drop can be related to the flow rate in the pipe. The manometric fluid is mercury (specific gravity = 13.6), and the specific gravity of the crude oil is 0.86. The reading obtained from the manometer is 18.7 cm. Compute the pressure differential.

Solution

This problem involves application of Equation (4.17):

$$P_a - P_b = H(\varrho_A - \varrho_B)g$$

where

$$H = 18.7 \text{ cm} \times \frac{1 \text{ m}}{100 \text{ cm}} = 0.187 \text{ m } (0.61 \text{ ft})$$

$$\varrho_A + 13.6 \times 1 \text{ g/cm}^3 \times \frac{\text{kg}}{10^3 \text{ g}} \times \frac{10^6 \text{ cm}^3}{\text{m}^3} = 13,600 \text{ kg/m}^3 \left(849 \frac{\text{lb}_m}{\text{ft}^3}\right)$$

$$\varrho_A = 860 \text{ kg/m}^3 \text{ (53.7 lb}_m/\text{ft}^3)$$

In SI units, the pressure differential is

$$P_a - P_b = 0.187 \text{ m}[(13,600 - 860)\text{kg/m}^3](9.807 \text{ m/s}^2)$$

$$= 23,364 \frac{\text{kg}}{\text{m-s}^2} = 23,364 \text{ N/m}^2$$

In English units, the pressure differential is

$$P_a - P_b = 0.61 \text{ ft}[(849 - 53.7)\text{lb}_m/\text{ft}^3] \frac{\left(32.2 \dfrac{\text{ft}}{\text{s}^2}\right)}{\left(32.174 \dfrac{\text{ft-lb}_m}{\text{lb}_f\text{-s}^2}\right)} \times \frac{\text{ft}^2}{144 \text{ in}^2}$$

$$= 3.37 \text{ psia}$$

Example 4.2

The system for the previous example is given in Figure 4.8. The pressure at point α is known to be P_a, N/m². Derive an expression for the pressure at point β.

Solution

The pressure at point 1 is

$$P_1 = P_a + \gamma_A(H + z)$$

where γ_A is the specific weight of fluid A.

The pressure at point 2 is the same as at point 1 because they are at the same level $(P_1 = P_2)$.

The pressure at point 3 is

$$P_3 = P_a + \gamma_A(z + H) - \gamma_B H$$

where γ_B is the specific weight of the manometric fluid.

The pressure at point β is, therefore,

$$P_\beta = P_3 - \gamma_A(z)$$

$$P_\beta = P_a + \gamma_A(z + H) - \gamma_B(H) - \gamma_A(z)$$

or, simply,

$$P_\beta = P_a + (\gamma_A - \gamma_B)H$$

We merely repeated our derivation of Equation (4.17). Hence, as stated earlier, the pressure differential $(P_\beta - P_a)$ is a function of the *difference* between the specific weights of the two fluids.

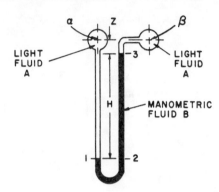

FIGURE 4.8. A differential manometer in operation for Example 4.2.

Example 4.3

A two-fluid manometer [Figure 4.4(c)] is being used to measure a point pressure in an exhaust duct where gas is flowing. One leg of the manometer is hooked up to the duct and the other is open to the atmosphere. The atmospheric pressure is 750 mm Hg. The two manometric fluids are water ($\gamma_w = 1$) and a light oil ($\gamma_o = 0.86$). The diameter of each of the large reservoirs is 25.4 mm and that of the small ones is 3.0 mm. The manometer reading is 150 mm. Determine the point pressure in the duct.

Solution

Examining Figure 4.4(c), a pressure balance for the U-tube is made:

$$P_a - P_b = (H - H_0)\left(\varrho_w - \varrho_o + \frac{A}{A'}\varrho_o\right)g$$

where

$H_0 =$ the manometer reading when $P_a = P_b$ (we can adjust the manometer reading to zero for no-flow in the duct)
$P_a =$ point pressure in the duct
$\varrho_w =$ density of the heavier fluid (water) $= 1000$ kg/m^3
$\varrho_o =$ density of the lighter fluid (oil) $= 860$ kg/m^3
$A' =$ cross-sectional area of each of the large reservoirs $= \frac{1}{4}\pi(25.4$ mm$)^2 \times 10^{-3}$ m/mm $= 0.51$ m^2
$A =$ cross-sectional area of each of the tubes forming the $U = 7.07 \times 10^{-3}$ m^2

Note that A/A' is relatively small and, hence, may be neglected. Also, because $H_0 = 0$, our equation reduces to Equation (4.17):

$$P_a - P_b = H(\varrho_w - \varrho_o)g$$

And solving for the pressure in the duct:

$P_a = P_b + H(\varrho_w - \varrho_o)g$
$P_a = (750 \text{ mm Hg})(1.3332 \times 10^2) + (0.150 \text{ m})[(1000 - 860) \text{ kg/m}^3] \times (9.8 \text{ m/s}^2)$
$P_a = 100.2 \text{ kN/m}^2$

PRESSURE FORCES ACTING ON SUBMERGED FLAT SURFACES

It is important to determine the pressure forces acting on the walls and floors of vessels when designing equipment. Excessive forces and pressure can warp and/or result in equipment failure. To evaluate such potential problems, an examination is needed of the physics of pressure forces. We shall begin the analysis by examining an elemental area of a vessel wall, $d\Omega$, submerged in the process fluid. By directing our attention to a small region, we can make some simplifying assumptions to the analysis, namely, that the surface under examination is flat and that the fluid pressure acting against it is distributed uniformly. The system under consideration is illustrated in Figure 4.9. We wish to estimate the magnitude of the resultant force on the area and determine the location of the center of pressure where the resultant force can be assumed to act.

Examining Figure 4.9, we take the coordinate axes locating the plane xoy on

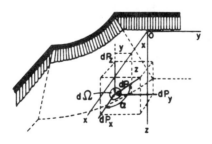

FIGURE 4.9. The coordinate system for analyzing the fluid pressure forces acting on an elemental area of a submerged vessel wall.

the fluid surface, with the oz axis in the downward direction. Then the pressure acting at the center of the elementary plane will be

$$P = \gamma z \qquad (4.19)$$

where z corresponds to the depth of the fluid from the free surface to the center of the plane. Hence, the resultant force on the area due to the fluid pressure is

$$dP = \gamma z d\Omega \qquad (4.20)$$

where dP is normal to the elemental plane.

We now expand the force dP in components dP_x, dP_y, dP_z parallel to their corresponding coordinate axes. So, for example,

$$dP_x = dP \cos\alpha = \gamma z d\Omega \cos\alpha$$

where α is the angle between the normal to the plane and the x-axis, and $d\Omega\cos\alpha$ is the projection of plane $d\Omega$ on the plane perpendicular to the ox-axis, i.e., on the plane zoy. We denote this projection as $d\Omega_{zy}$. Then,

$$dP_x = \gamma z d\Omega_{zy} \qquad (4.21)$$

In the same manner, the components dP_y and dP_z may be presented as follows:

$$dP_y = \gamma z d\Omega_{xz} \qquad (4.22)$$

$$dP_z = \gamma z d\Omega_{yx} \qquad (4.23)$$

The RHS of Equation (4.23) is the weight of an elementary fluid column having a cross section equivalent to the horizontal projection of a given flat element of a wall whose height is the depth of submergence of the element's center of gravity. Because the element $d\Omega$ is infinitesimal, the product $\gamma z d\Omega_{yx}$ may be equated to the weight of fluid column acting on the elemental flat wall.

Suppose we wish to determine the pressure acting against an inclined plane at some angle α, as shown in Figure 4.10. The ox-axis is directed along the free surface of the fluid to intersect with a wall and the oz-axis is vertically downward. To determine the total force P on the wall, we must sum the forces acting on the wall elements, i.e.,

$$P = \gamma \int_\Omega z d\Omega \qquad (4.24)$$

Pressure Forces Acting on Submerged Flat Surfaces

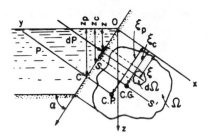

FIGURE 4.10. The coordinate system for analyzing the fluid pressure forces acting on an inclined vessel wall.

Denoting ξ as the distance from the ox-axis to the center of gravity of $d\Omega$, we have

$$\xi = \frac{z}{\sin \alpha}$$

Accounting for this in Equation (4.20), we obtain

$$P = \gamma \sin \alpha \int_\Omega \xi d\Omega$$

The integral $\int_\Omega \xi d\Omega = S_x(\Omega)$ is the static moment S_x of the area Ω with respect to the ox-axis, which is equal to the product of the area and distance ξ_c from the center of gravity. Thus,

$$S_x(\Omega) = \Omega \xi_c$$

and

$$P = \gamma \Omega \xi_c \sin \alpha \qquad (4.25)$$

However, we must note the following:

$$P = \gamma \Omega z_c = P_c \Omega \qquad (4.26)$$

where $z_c = \xi_c \sin \alpha$ is the submergence depth of the center of gravity and $P_c = \gamma z_c$ is the excessive pressure at the center of gravity of the wall. The component of pressure along the oz-axis may be expressed as follows:

$$P_z = \gamma \int_\Omega z d\Omega_{xy} \qquad (4.27)$$

Because $\int_\Omega z d\Omega_{xy}$ is the volume of the fluid over a flat wall, the vertical component of the total pressure is equivalent to the weight of this volume of liquid.

We will now determine the position of point C (the center of pressure) acting on the wall through which the line of action of total pressure P passes. The coordinate of the center of pressure (Figure 4.10) is ξ_p. From a principle of statics, the moment of the resultant of a force system is equal to the sum of the components of the forces. Taking momentums of forces with respect to the ox-axis, we obtain

$$\xi_p P = \int_\Omega \xi dP \qquad (4.28)$$

And, accounting for Equations (4.20) and (4.25), we obtain the following:

$$\xi_p \xi_c \gamma \Omega \sin \alpha = \int_\Omega \xi^2 \sin \alpha d\Omega$$

or

$$\xi_p \xi_c \Omega = \int_\Omega \xi^2 d\Omega$$

The value of $\int_\Omega \xi^2 d\Omega = J_x$ is the moment of inertia of the wall area Ω about the ox-axis. J_x is basically the product of area Ω and the square of inertia radius ϱ_x about the ox-axis:

$$\xi_p \xi_c \Omega = \varrho_x^2 \Omega$$

Hence,

$$\xi_p = \frac{\varrho_x^2}{\xi_c} \qquad (4.29)$$

Trace the line SS' through the center of gravity of the wall parallel to the ox-axis. In accordance wih a known relationship between the momentum of inertia of an area about parallel axes, the moment of inertia J_x may be written in the following form:

$$J_x = \xi_c^2 \Omega + J_s \qquad (4.30)$$

where J_s is the moment of inertia of the wetted wall area about the SS'-axis. Therefore,

$$\varrho_x^2 \Omega = \xi_c^2 \Omega + \varrho_s^2 S\Omega \qquad (4.31)$$

where ϱ_s is the radius of inertia of the wall about the SS'-axis. Thus, Equation (4.29) has the following form:

$$\xi_p \xi_c = \xi_c^2 + \varrho_s^2 S \qquad (4.32)$$

Hence,

$$\xi_p = \xi_c + \frac{\varrho_s^2 S}{\xi_c} \qquad (4.33)$$

Keeping in mind that all terms in Equation (4.33) are positive, we must conclude that the center of pressure for a flat inclined wall is always located below the center of gravity. The following example problems apply the above principles.

Example 4.4

An 18-ft-wide, 35-ft-high crude oil storage tank is vented to the atmosphere. It is believed that water accumulation at the tank bottom may have seriously reduced the available storage volume, as well as promoting corrosion of the tank shell. If the gauge pressure at the tank bottom is 13.9 psig, determine the volumes of oil and water in barrels. The specific gravity of the crude is 0.91.

Solution

As the tank is vented to the atmosphere, we can assume that the pressure over the oil surface is that of the atmosphere, i.e., $P_t = 14.696$ psia. From a pressure balance, the pressure at the bottom of the tank is

$$P_b = h_o \varrho_o \frac{g}{g_c} + P_t + h_w \varrho_w \frac{g}{g_c}$$

where ϱ_o, ϱ_w are the densities of oil and water, respectively, h_o is the height of the oil, and h_w is the height of the water layer in the tank. Assuming the tank is filled to capacity, the overall tank height is

$$H_T = h_o + h_w$$

or the oil height is

$$h_o = H_T - h_w = 35 - h_w$$

Substituting for h_o in our pressure balance and noting that the gauge pressure at the bottom is equal to the absolute pressure P_b minus the atmospheric pressure $P_{gauge} = P_b - P_t$,

$$P_b - P_t = (H_T - h_w)\varrho_o \frac{g}{g_c} + h_w \varrho_w \frac{g}{g_c}$$

$$P_{gauge} = H_T \varrho_o \frac{g}{g_c} + h_w \frac{g}{g_c}(\varrho_w - \varrho_o)$$

$$13.9 \text{ lb}_f/\text{in}^2 \times 144 \text{ in}^2/\text{ft}^2 = (35 \text{ ft})(0.91 \times 62.4 \text{ lb}_m/\text{ft}^3)\left(1.0 \frac{\text{lb}_f}{\text{lb}_m}\right)$$

$$+ h_w \left(1.0 \frac{\text{lb}_f}{\text{lb}_m}\right)[(1.0 - 0.91) \times 62.4 \text{ lb}_m/\text{ft}^3]$$

$$2001.6 = 1987.4 + 5.62 h_w$$

$$h_w = 2.5 \text{ ft}$$

and the height of the oil is $h_o = 35 - 2.5 = 32.5$ ft. The cross-sectional area of the tank is $\frac{1}{4}\pi(18 \text{ ft})^2 = 254.5 \text{ ft}^2$. The volume of oil is

$$V_o = 32.5 \text{ ft} \times 254.5 \text{ ft}^2 \times 7.48 \frac{\text{gal}}{\text{ft}^3} \times \frac{\text{bbl}}{42 \text{ gal}} = 1473 \text{ bbl}$$

and that of the water is

$$V_w = 2.5 \times 254.5 \times \frac{7.48}{42} = 113.3 \text{ bbl}$$

Hence, about 7% of the tank volume has been lost to water accumulation.

Example 4.5

A reactor vessel is fitted with a 6-in.-diameter circular glass viewing port on an inclined wall ($\theta = 62°$). The system is illustrated in Figure 4.11. The centroid of the viewing port is at a depth of 3.5 ft from the liquid interface. The specific gravity of the liquid mixture is approximately 0.79.

1 Evaluate the magnitude of the resultant fluid force on the viewing port.
2 Determine the location of the center of pressure.

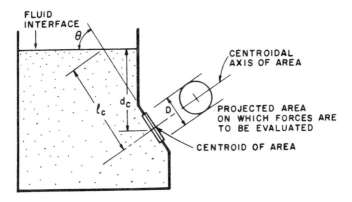

FIGURE 4.11. Defines system coordinates for Example 4.5.

Solution

Part 1: The area of interest is the circular viewing port. For the simple area of a circle, the centroid exists at its center. From the problem description, we know the vertical depth from the fluid interface to the centroid of the viewing port $d_c = 3.5$ ft. Note that ℓ_c and d_c in Figure 4.11 are related through angle θ as follows:

$$\sin\theta = d_c/\ell_c$$

Hence,

$$\ell_c = d_c/\sin\theta = 3.5 \text{ ft}/\sin 62° = 3.96 \text{ ft}$$

ℓ_c and d_c will be needed later on. The area of the viewing port is

$$A = \frac{1}{4}\pi(6/12)^2 = 0.20 \text{ ft}^2$$

The resultant force, F_R, acting on the port is

$$F_R = \gamma d_c A \frac{g}{g_c}$$

where γ is the specific weight of the liquid:

$$\gamma = 0.79 \times 62.4 \text{ lb}_m/\text{ft}^3 = 49.3 \text{ lb}_m/\text{ft}^3$$

Thus,

$$F_R = 49.3 \ \frac{\text{lb}}{\text{ft}^3} \times 3.5 \text{ ft} \times 0.20 \text{ ft}^2 = 34.5 \text{ lb}_f$$

Part 2: For a circle, the area moment of inertia is given by the following formula:

$$J = \frac{\pi D^4}{64}$$

or

$$J = \frac{\pi}{64} (6/12)^4 = 3.07 \times 10^{-3} \text{ ft}^4$$

And we know that $\ell_c = 3.96$ ft and $A = 0.20$ ft². The distance normal from the resultant force vector to the fluid interface is

$$x = \ell_c + \frac{J}{\ell_c A} = 3.96 + \frac{3.07 \times 10^{-3} \text{ ft}^4}{(3.96 \text{ ft})(0.20 \text{ ft}^2)}$$

$$x = 3.96 \text{ ft} + 3.88 \times 10^{-3} \text{ ft} = 3.964 \text{ ft}$$

which means that the center of pressure will occur essentially at the centroid of the viewing port.

Example 4.6

The cross section of a vessel is in the form of a parabolic segment, as shown in Figure 4.12. Determine the resultant fluid pressure acting on the vessel's vertical front wall for a fluid depth H and fluid interface width $2a$.

Solution

The area of the parabolic segment is as follows:

$$\Omega = \frac{2}{3} aH$$

The distance from the center of gravity to the apex of a parabola is equal to $3/5H$; consequently,

$$z_c = H - \frac{3}{5}H = \frac{2}{5}H$$

The moment of inertia of the segment about the boundary chord is

$$J_x(\Omega) = \frac{8H^2}{35}\Omega$$

Thus, the square of inertia radius is

$$\varrho_x^2 = \frac{8H^2}{35}$$

and, from Equation (4.26), we obtain the following:

$$P = \gamma z_c \Omega = \gamma \frac{2}{5} H \frac{2}{3} aH = \frac{4}{15}\gamma aH^2$$

The coordinate of the center of pressure can be obtained from Equation (4.29). Hence, the answer is

$$z_p = \frac{8H^2}{35} \Big/ \frac{2H}{5} = \frac{4}{7}H$$

Example 4.7

A conical vessel with no bottom is mounted on a horizontal surface. The dimensions of the vessel are given in Figure 4.13. The specific weight of the vessels shell is γ_m. Determine the liquid height that will cause the vessel to tear away from the horizontal surface.

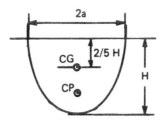

FIGURE 4.12. Parabolic cross section of a vessel for Example 4.6. CG and CP denote the center of gravity and the center of the pressure force, respectively.

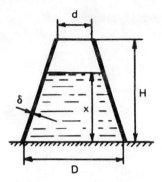

FIGURE 4.13. Conical vessel with no bottom mounted on a horizontal surface defining the system for Example 4.7.

Solution

The weight of the vessel is

$$G_v = \gamma_m \pi \delta \frac{D + d}{4} \sqrt{4H^2 + (D - d)^2}$$

The force that will tear the vessel away from the horizontal surface will result from the liquid pressure acting on one of the vessel's side walls. This force is directed vertically upward and is equal to the difference between the liquid weight in the cylinder and that in the cone, where

$$V_c = \frac{\pi D^2}{4} x$$

and

$$V_{cone} = \frac{\pi x}{12}\left[3D^2 - \frac{3D}{H}x(D - d) + \frac{x^2}{H^2}(D - d)^2\right]$$

The tearing force is equal to the following:

$$G_\ell = \gamma_\ell \frac{\pi D^2}{12} \times \left(1 - \frac{d}{D}\right)\left[\frac{3x}{H} - \frac{x^2}{H^2}\left(1 - \frac{d}{D}\right)\right]$$

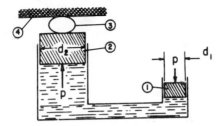

FIGURE 4.14. A scheme for a hydraulic press.

Equating the above expression to the weight of the vessel, we obtain an equation for liquid level x raised to the third power:

$$\gamma_\ell \frac{\pi D^2}{12} \frac{x^x}{H} \left(1 - \frac{d}{D}\right) \left[3 - \frac{x}{H}\left(1 - \frac{d}{D}\right)\right]$$

$$= \gamma_m \frac{\pi \delta (D + d)}{4} \sqrt{4H^2 + (D - d)^2}$$

This equation can have only one real root if

$$\left| \frac{\gamma_m}{\gamma_\ell} \frac{\delta(D^2 - d^2) \sqrt{4H^2 + (D - d)^2}}{D^3 H} \right| < \frac{4}{3}$$

HYDROSTATIC MACHINES

The design and operation of hydrostatic machines is based on the application of the basic hydrostatic equation. One example of a hydrostatic machine is illustrated in Figure 4.14. The system shown is a hydraulic press, which can be used for pressing and forming briquets of different materials. If a relatively small force is applied to piston 1, which moves in a cylinder of diameter d_1, a pressure P is established on the piston. In accordance with Pascal's law, the same pressure will be exerted on piston 2 in the large-diameter cylinder d_2. The force acting on piston 1 is

$$P_1 = p \frac{\pi d_1^2}{4} \qquad (4.34)$$

and the force acting on piston 2 is

$$P_2 = p \frac{\pi d_2^2}{4} \qquad (4.35)$$

As a result, the piston in the cylinder of greater diameter will transfer the force, which is many times greater than the force applied to the piston in cylinder of diameter d_1 to compressing material 3. The forces applied to compressing material 3 increase proportionally with the ratio of the diameters of the piston cylinders. Thus, by harnessing relatively small forces, it is possible to press material 3 located between piston 2 and the immovable plate 4.

Further discussions of hydrostatic machines and other principles described in this chapter may be found in the references listed.

NOTATION

A = cross-sectional area, m²
CG = center of gravity
CP = center of pressure
D = diameter, m
d_c = vertical depth from liquid surface, m
F_R = resultant force, N
g = gravitational acceleration, m/s²
g_c = conversion factor, 32.174 ft-lb$_m$/lb$_f$-s²
H = fluid displacement or height, m
h = height, m
J = moment of inertia, m⁴
J_s = moment of inertia of wetted surface, m⁴
ℓ_c = distance normal from center of gravity to fluid surface, m
P, p = pressure, N/m²
RHS = right-hand side
S = static moment of area, m³
V = volume, m³
z = level, m
Z_c = submergence depth to center of gravity, m

Greek Symbols

α = angle between normal to plane and x-axis, °
γ = specific gravity
θ = angle, °
ξ_c = coordinate of center of gravity, m
ξ_p = coordinate of center of pressure, m
ϱ_s = radius of inertia of wall, m
ϱ_x = inertia radius, m
Ω = area, m²

REFERENCES

CHEREMISINOFF, N. P. *Applied Fluid Flow Measurement: Fundamentals and Technology.* New York:Marcel Dekker Inc. (1979).

CHEREMISINOFF, N. P. *Process Level Instrumentation and Control.* New York:Marcel Dekker Inc. (1981).

BIBLIOGRAPHY

ADDISON, H. *Hydraulic Measurements.* New York:John Wiley & Sons, Inc. (1949).

DIEDERICH, H. and W. Andrae. *Experimental Mechanical Engineering, Vol. 1.* New York:John Wiley & Sons, Inc. (1930).

PERRY, R. H. and C. H. Chilton, eds. *Chemical Engineer's Handbook,* 5th ed. New York:McGraw-Hill Book Co. (1973).

STREETER, V. L. *Handbook of Fluid Dynamics.* New York:McGraw-Hill Book Co. (1961).

CHAPTER 5

Hydrodynamics: Single-Fluid Flows

INTRODUCTION

An understanding of the basic physics governing flow dynamics is essential to develop governing equations to which we refer as hydrodynamic laws. It is the proper application of these hydrodynamic laws that enables a multitude of unit operations to be designed and operated efficiently and safely throughout the process industries. Indeed, it is not possible to design rationally for heat and mass transfers without properly defining the hydrodynamics of the flow system.

The principles presented in this chapter are the foundation for the study of unit operations. Presented here are the governing equations describing the motion of single-phase fluids, which establish a basis for describing more complex multiphase flows encountered in many unit operations. Emphasis will be placed on the analysis of pipe flow for both compressible and incompressible systems. In addition, an analysis is included of the important system of film flow. The motion of single fluids through piping and equipment constitutes the first class of problems of system analysis encountered in unit operations, namely, the internal problems of hydrodynamics.

CHARACTERISTICS OF FLUIDS IN MOTION

Steady-State versus Transient Flows

Before addressing the complex behavior of fluids in motion and their subsequent effects on the transfers of heat and mass in various unit operations, an understanding is needed of the physics involved. Therefore, we will begin our discussions by describing certain characteristics common to all fluid behavior, regardless of the particular unit operation being performed. It is important, especially for later discussions, that the characteristics and definitions presented in this chapter be clearly understood. To begin, a distinction must be made between steady-state and unsteady-state, or transient flows.

In a process or operation described as being a steady state, the flow rates and properties of the fluid (e.g., temperature, pressure, composition, density, and velocity) at each point within the system are constant with respect to time. These properties may vary from point to point within the system, but at any one location at any point in time they remain unchanged. Velocity, for example, may have different values at different points [$v_x = f(x,y,z)$], but at any one point it does not change with time, i.e., $\partial v_x/\partial t = 0$. Hence, for steady-state flows only special variations may exist.

In contrast, for systems that have *transient flows,* the properties that influence fluid motion *change with respect to time.* As an example, for a fluid flowing through a pipe in the x-direction, the velocity at any one location within the pipe is not only a function of the spacial coordinates x, y, and z, but also of the time t, i.e., $v_x = f(x,y,z,t)$; consequently, $\partial v_x/\partial t \neq 0$. A specific example of a transient system is discharge through an orifice on a tank operating at variable liquid level. As the level in the tank decreases, so does the discharge velocity. Steady-state flow processes are typical in unit operations, whereas many transient flows are common to batch operations or are associated with startups and shutdowns. Transient flows are also associated with upset conditions in process operation flows. In continuous operations, transient conditions often result from flow regimes caused by phase changes or flow transitions at various points in the system.

To characterize an unsteady-state flow process with respect to time, we will analyze the flow properties at the fixed points, i.e., we shall examine changes with time at constant spacial coordinates. For every moving fluid particle, the change of its parameters in time and space will not be described by a partial derivative but by a total derivative. This total derivative is called a *substantial derivative* or, more commonly, a *Lagrangian derivative.*

Let u be a dummy variable that denotes any property of the flow that changes with respect to time and space (e.g., temperature, pressure, density, concentration, or any component of velocity w_x, w_y, and w_z). Suppose we observe the flow and can measure the fluctuations of u at any moment at a fixed point in space (x,y,z). As stationary observers, we would note that the rate of change in u with time is constant. The rate of change in u may be described by a partial derivative $\partial u/\partial t$ and, hence, the variation in u over a time interval dt is $\partial u/\partial t\, dt$. This value is the *local* variation of the given variable and, at steady state, is equal to zero.

If we now move along with the flow, then the measured value will be the sum of two components, that is, u is a resultant vector. Assume a fluid particle is displaced from point $A(x,y,z)$ to point B [$(x + dx)$, $(y + dy)$, $(z + dz)$] over some time dt. As a result of this displacement, the corresponding path of projections dx, dy, dz becomes $\partial u/\partial x\, dx$, $\partial u/\partial y\, dy$, and $\partial u/\partial z\, dz$. Note that these quantities are not taken with respect to time. Thus, in the case of steady-state flow (where there are no local changes), u changes by a value resulting from the displacement from A to B:

$$du = \frac{\partial u}{\partial x} dx + \frac{\partial u}{\partial y} dy + \frac{\partial u}{\partial z} dz \qquad (5.1)$$

Equation (5.1) characterizes the *convective* change of parameter u.

For unsteady-state flow, where $u = f(x,y,z,t)$, u changes over a time interval dt by a value $\partial u/\partial t\, dt$. Consequently, the *total* change in u is *the sum of the local and convective variations*:

$$du = \frac{\partial u}{\partial t} dx + \frac{\partial u}{\partial x} dy + \frac{\partial u}{\partial y} dy + \frac{\partial u}{\partial z} dz \qquad (5.2A)$$

or

$$\frac{du}{dt} = \frac{\partial u}{\partial t} + \frac{\partial u}{\partial x}\frac{dx}{dt} + \frac{\partial u}{\partial y}\frac{dy}{dt} + \frac{\partial u}{\partial z}\frac{dz}{dt} \qquad (5.2B)$$

Assuming u to be fluid velocity, we note that the values $dx/dt = w_x$, $dy/dt = w_y$, $dz/dt = w_z$ are components on the corresponding coordinate axis, and hence, we obtain the following:

$$\frac{du}{dt} = \frac{\partial u}{\partial t} + \frac{\partial u}{\partial x} w_x + \frac{\partial u}{\partial y} w_y + \frac{\partial u}{\partial z} w_z \qquad (5.3A)$$

When $\partial u/\partial t = 0$ (steady state),

$$\frac{du}{dt} = \frac{\partial u}{\partial x} w_x + \frac{\partial u}{\partial y} w_y + \frac{\partial u}{\partial z} w_z \qquad (5.3B)$$

Equations (5.3A) and (5.3B) represent the substantial derivative of the given parameter, which characterizes changes in both time and space. We shall make the analysis more specific by examining a *continuous fluid flow* situation. Consider a flow in which the same amount of fluid enters and leaves a system, i.e., there is no accumulation or depletion of available fluid supply over some arbitrary time interval. To analyze this system, we must define the terms fluid mass velocity and linear velocity.

Fluid Mass and Linear Velocities

Let us first consider a very simple case of flow through a conduit of constant cross-sectional area. The amount of fluid flowing through a specified cross section is referred to as the *volumetric flow rate*, or *mass (weight) flow rate*, having units of m³/s or kg/s, respectively.

From the mass rate W(kg/hr) and cross-sectional area F(m²), a *superficial mass velocity* can be computed. The superficial mass velocity is defined as the ratio of total mass flow rate divided by the total area of flow; i.e., in the case of a circular conduit, $F = \frac{1}{4}\pi D^2$,

$$G = \frac{W}{F}, \text{ kg/m}^2\cdot\text{hr} \tag{5.4}$$

From the fluid density ϱ(kg/m³) and mass rate W(kg/s), the volumetric flow rate is determined as follows:

$$V = \frac{W}{\varrho}, \text{ m}^3/\text{hr} \tag{5.5}$$

Volumetric flow rate is defined simply as the volume of fluid flowing through a given cross section per unit time. The ratio of volumetric rate to the cross section is the definition of the mean linear velocity of the flow:

$$\overline{W} = \frac{V}{F}, \text{ m/hr} \tag{5.6}$$

The local linear fluid velocity, i.e., the velocity at a given point in the flow, depends strongly on whether the fluid is under the influence of solid boundaries. Fluid elements close to the pipe axis move faster than those in the region of the wall. Hence, to compute the volumetric flow rate, information on a local velocity, for example, along the pipe axis, is insufficient. We require information on how the velocity changes over the cross section. For a symmetrical velocity distribution about the pipe axis, w is a function of distance perpendicular from the axis to the pipe wall:

$$w = f(r) \tag{5.7}$$

Let us consider an elemental circular element of the flow having radius r and thickness dr, moving along a pipe of radius R. A front view of this system is illustrated in Figure 5.1. The perimeter of this elemental ring is $2\pi r$ and its area is

$$dF = 2\pi r dr$$

FIGURE 5.1. Front view of fluid flowing through a pipe.

If the linear velocity of the ring is w, i.e., the local velocity, then, following Equation (5.6), the local volumetric flow is

$$dV = w2\pi r dr$$

Hence, the volumetric rate over the total cross section is as follows:

$$V = 2\pi \int_0^R wr\,dr \qquad (5.8)$$

Once the relationship between local linear velocity w and distance normal from the pipe axis to its walls is known, the integral in Equation (5.8) can be evaluated for the volumetric flow rate. The integral may be evaluated either exactly or graphically.

The average linear velocity may be evaluated similarly. For a circular flow section, $F = \pi r^2$, using Equations (5.6) and (5.8), the following is obtained:

$$\overline{w} = \int_0^1 w\,d\left(\frac{r}{R}\right)^2 \qquad (5.9)$$

The local velocity w in Equation (5.8) can be rewritten in terms of the fraction of the pipe radius r/R. Knowing $w = f/(r)$, it is also possible to determine $w = f(r/R)$ and $w = f(r/R)^2$. By integrating this relationship, graphically, for example, over the limits of r/R from 0 to 1, the average velocity of the flow is obtained. From Equations (5.4)–(5.6), several other useful relations may be written. The mass velocity can be expressed as the average velocity and specific weight of the fluid:

$$G = \overline{W}\gamma \qquad (5.10A)$$

The mass flow rate may be expressed as

$$W = \overline{W}\gamma F \qquad (5.10B)$$

Now we are ready to consider the continuous flow system described at the end of the previous subsection. According to our definition of continuous flow, the mass rate must be the same at any section within the system. That is,

$$W_1 = W_2 = W_3 = \ldots \tag{5.11A}$$

Starting with Equation (5.4), the continuity of flow may be stated as follows:

$$G_1 F_1 = G_2 F_2 = G_3 F_3 = \ldots \tag{5.11B}$$

And, from Equation (5.10B),

$$\overline{W}_1 \gamma_1 F_1 = \overline{W}_2 \gamma_2 F_2 = \overline{W}_3 \gamma_3 F_3 = \ldots \tag{5.12}$$

Equations (5.11A,B) and (5.12) represent the material balance for the flow system. Equation (5.12) has special relevance to compressible flows. In gas flow, pressure, temperature, and specific gravity may vary at different cross sections and, hence, different linear velocities will result. Equation (5.12) permits us to establish the relationship between these velocities (when cross sections are changed) and their specific gravities. For liquid flow in a pipe of constant cross-sectional area, i.e., $F = F_1 = F_2 = F_3$, the balance equation is considerably simplified; and, from Equations (5.11B) and (5.12), we obtain

$$G_1 = G_2 = G_3 = \ldots = \overline{W}_1 \varrho_1 = \overline{W}_2 \varrho_2 = \overline{W}_3 \varrho_3 \tag{5.13}$$

Equation (5.13) states that for any cross section of piping the mass rates of gas or liquid will be the same even if the temperatures and pressures undergo considerable changes. Applying Equation (5.12) to a liquid flow undergoing a small temperature change, and even at large pressure changes, we may assume that the specific density of the liquid is approximately constant. Consequently, the equation of continuity for liquid flow is

$$\overline{W}_1 F_1 = \overline{W}_2 F_2 = \overline{W}_3 F_3 \tag{5.14A}$$

And, from Equation (5.6), the volumetric rate is

$$V_1 = V_2 = V_3 = \ldots \tag{5.14B}$$

According to the above, the average linear velocities for liquid flow are inversely proportional to the cross-sectional areas of flow and, hence, the volumetric rates in different sections are the same. For piping of constant cross section, the linear velocities are therefore the same:

$$\overline{W}_1 = \overline{W}_2 = \overline{W}_3 = \ldots \tag{5.15}$$

Equations (5.13), (5.14), and (5.15) may be used in practical calculations for liquids and, with some approximations, for gases in cases of small temperature and pressure changes, i.e., when the specific weight of the gas can be considered constant.

Residence Time

To determine *liquid residence time* in a system (e.g., in a pipe or piece of equipment), assume that the mass rate W does not change over the cross section and that there are no stagnant regions in the system. If dx is the distance along the direction of the flow and if the flow travels over dx in time dt, then the average linear velocity is simply

$$\overline{W} = \frac{dx}{dt} \qquad (5.16)$$

From Equation (5.10B), we have

$$W = \gamma F \frac{dx}{dt} \qquad (5.17)$$

where F is the cross-sectional area normal to the direction of flow. Note that the elemental volume of the pipe or piece of equipment is $dV_0 = Fdx$. When γ is constant, on integration of Equation (5.17) the residence time is obtained:

$$t = \frac{\gamma}{W} V_0 = \frac{V_0}{V} \qquad (5.18)$$

Hence, residence time depends only on the system volume V_0, and the volumetric flow. Equation (5.18) states in simple terms that residence time is the ratio of system volume to the volumetric flow rate.

Further discussions of the principles presented in this subsection are given by Bird et al. (1960), Bennet and Myers (1964), Curle and Davies (1964), and Streeter and Wylie (1979). The following two example problems apply some of the principles described; additional problems may be found at the end of this chapter.

Example 5.1

Oil (0.85 specific gravity) is flowing through two sections of piping that vary in cross section. The average velocity in section 1 is 1.2 m/s. The diameter of pipe section 1 is 8.9 cm and that of section 2 is 3.8 cm. Determine the following:

1. The velocity at smaller section 2
2. The volumetric flow rate
3. The mass flow rate
4. The mass velocity in each section

Solution

1. The velocity at the smaller section is obtained from continuity Equation (5.14A):

$$\overline{W}_1 F_1 = \overline{W}_2 F_2$$

$$\overline{W}_2 = \overline{W}_1 \cdot \frac{F_1}{F_1}$$

$$F_1 = \frac{\pi}{4} D_1^2 = \frac{\pi}{4} (0.089 \text{ m})^2 = 6.221 \times 10^{-3} \text{ m}^2$$

$$F_2 = \frac{\pi}{4} D_2^2 = \frac{\pi}{4} (0.038 \text{ m})^2 = 1.134 \times 10^{-3} \text{ m}^2$$

Hence,

$$\overline{W}_2 = \left(1.2 \, \frac{\text{m}}{\text{s}}\right) \left(\frac{6.221 \times 10^{-3} \text{ m}^2}{1.134 \times 10^{-3} \text{ m}^2}\right) = 6.6 \, \frac{\text{m}}{\text{s}}$$

For the steady flow of a liquid, as the flow area is decreased the average velocity increases. This answer is independent of pressure and elevation changes.

2. To obtain the volumetric flow rate, we again use the principle of continuity, i.e., we can use the conditions either at section 1 or at section 2 to compute V [see Equation (5.14B)]:

$$V = F_1 \overline{W}_1 = (6.221 \times 10^{-3} \text{ m}^2)(1.2 \text{ m/s}) = 7.5 \times 10^{-3} \, \frac{\text{m}^3}{\text{s}}$$

or

$$V = F_2 \overline{W}_2 = (1.134 \times 10^{-3} \text{ m}^2)(6.6 \text{ m/s}) = 7.5 \times 10^{-3} \, \frac{\text{m}^3}{\text{s}}$$

3. The mass flow rate is simply:

$$W = V\gamma$$

$$= \left(7.5 \times 10^{-3} \frac{m^3}{s}\right)\left(850 \frac{kg}{m^3}\right) = 6.38 \text{ kg/s}$$

4. The mass velocity is defined by Equation (5.10A):

In section 1,

$$G_1 = \overline{W}_1 \varrho = \left(1.2 \frac{m}{s}\right)\left(850 \frac{kg}{m^3}\right) = 1020 \frac{kg}{m^2\text{-}s}$$

In section 2,

$$G_2 = \overline{W}_2 \varrho = \left(6.6 \frac{m}{s}\right)\left(850 \frac{kg}{m^3}\right) = 5610 \frac{kg}{m^2\text{-}s}$$

We have invoked the assumption that the oil's specific gravity did not change as it passed from one section to the other.

Example 5.2

The exhaust portion of a room's ventilation system consists of a 12 in × 4 in rectangular withdrawal duct. The air is exhausted on the roof of the building through a circular stack with an inside diameter of 16 in. The mean temperature of the air in the room is 85°F at 14.7 psia, and the average velocity at the withdrawal duct is 1500 fpm. Neglecting temperature fluctuations in the exhaust ductwork and outside the building, determine the following:

1 The density of the air in the circular exhaust stack for a measured average velocity of 700 fpm
2 The mass flow rate of air exhausted

Solution

1. The density of air at STP is approximately 0.0808 lb/ft³. From the ideal gas law, Equation (3.7), the density of air at 85°F is

$$\varrho = \varrho_0 \frac{T_0}{T} = \left(0.0808 \frac{lb}{ft^3}\right)\left(\frac{32 + 460°F}{85 + 460°F}\right) = 0.0729 \frac{lb}{ft^3}$$

According to the continuity equation for gas flow, Equation (5.12),

$$\overline{W}_1 \varrho_1 F_1 = \overline{W}_2 \varrho_2 F_2$$

$$\varrho_2 = \varrho_1 \left(\frac{F_1}{F_2}\right)\left(\frac{\overline{W}_1}{\overline{W}_2}\right)$$

$$F_1 = 12 \text{ in} \times 4 \text{ in} = 48 \text{ in}^2$$

$$F_2 = \frac{\pi}{4} (16 \text{ in})^2 = 201 \text{ in}^2$$

$$\varrho_2 = 0.0729 \frac{\text{lb}}{\text{ft}^3} \left(\frac{48 \text{ in}^2}{201 \text{ in}^2}\right)\left(\frac{1500 \text{ fpm}}{700 \text{ fpm}}\right) = 0.0373 \frac{\text{lb}}{\text{ft}^3}$$

2. The mass flow rate of air exhausted is

$$W = \gamma_1 F_1 \overline{W}_1 = \gamma_2 F_2 \overline{W}_2$$

$$W = 0.0729 \frac{\text{lb}}{\text{ft}^3} \times 48 \text{ in}^2 \times 1500 \frac{\text{ft}}{\text{min}} \times \frac{\text{ft}^2}{144 \text{ in}^2} \times \frac{60 \text{ min}}{\text{hr}}$$

$$W = 2187 \frac{\text{lb}}{\text{hr}}$$

REGIMES OF FLOW

Laminar versus Turbulent Flows

Two general regimes of flow describe the nature of fluid motion and the interactions between fluid particles. These regimes are termed laminar and turbulent flows. The *laminar regime* occurs at relatively low fluid velocities. In this regime, the flow may be visualized as layers that slide over each other providing smooth flow patterns. No macroscopic mixing of fluid particles occurs in this regime. In the *turbulent regime,* fluid velocities are higher, and an unstable pattern with the bulk flow is observed in which eddies, or small packets of fluid particles, move at all angles to the axial line of normal total flow. However, a thin layer exists near the wall where the fluid motion is still laminar. This region is known as the *laminar boundary layer.*

Classical studies by Reynolds in 1883 showed that the transition from laminar to turbulent flow in tubes is a function of the fluid velocity, density, and viscosity, and the tube diameter. The dependency of flow regime on these variables is summarized by a dimensionless group known as the *Reynolds number:*

$$Re = \frac{wD}{\nu} = \frac{wD\varrho}{\mu} = \frac{WD}{\mu} \qquad (5.19)$$

where

ν = kinematic viscosity
ϱ = density
μ = dynamic viscosity
w = fluid velocity
W = mass velocity
D = tube diameter (or some characteristic size of the system through which flow occurs)

Equation (5.19) indicates that turbulent motion increases with tube diameter, fluid velocity, and density, or with decreasing viscosity. The value of the Reynolds number corresponding to the transition from one regime to another is referred to as the *critical Reynolds number,* or simply, the *critical value.*

In a straight circular pipe, when the Reynolds number is less than 2100 the flow is considered laminar. When the Reynolds number exceeds 4000 the flow is turbulent, except for some very special cases. The flow between these two values is referred to as the *transition region,* where the motion may be either laminar or turbulent.

Consider steady-state laminar flow of a fluid of constant density ϱ through a tube of length L and radius R. In the analysis to follow, "end effects" shall be ignored, i.e., we will ignore the fact that at the tube's entrance and exit the flow will not be parallel to the tube axis. Visualize a portion of the flow as a cylindrical layer of length ℓ and radius r. The system is illustrated in Figure 5.2. The motion of the layer results from a difference in pressure forces P_1 and P_2 at both ends of the fluid cylinder:

$$P_1 - P_2 = (p_1 - p_2)\pi r^2 \qquad (5.20)$$

where p_1 and p_2 are the hydrostatic pressures at sections 1 and 2 (Figure 5.2).

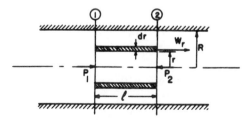

FIGURE 5.2. Plan view of fluid at laminar flow through a tube.

The motion of the fluid cylinder is resisted by a friction force

$$T = -\mu F \frac{dw_r}{dr} \qquad (5.21)$$

where

w_r = fluid velocity along the x-axis at a distance r from the tube centerline
$F = 2\pi r \ell$ = the cylinder's outside surface area
μ = fluid viscosity

The minus sign in Equation (5.21) denotes that the fluid velocity decreases and radius r increases. That is, as r approaches the radius of the tube w_r becomes smaller, and at $r = R$, $w_r = 0$. The pressure gradient $P_1 - P_2$ represents the driving force for flow and must be sufficiently large to overcome frictional force T, i.e.,

or

$$\left. \begin{array}{c} (P_1 - P_2)\pi r^2 = -\mu 2\pi r \ell \dfrac{dw_r}{dr} \\[1em] \dfrac{P_1 - P_2}{2\mu \ell} r\, dr = -dw_r \end{array} \right\} \qquad (5.22)$$

This differential equation may be integrated over the limits of $r = 0$ to R and $w = w_r$ to 0:

$$\int_r^R \frac{P_1 - P_2}{2\mu \ell} r\, dr = -\int_{w_r}^0 dw_r \qquad (5.23)$$

Hence,

$$\frac{P_1 - P_2}{2\mu \ell}\left(\frac{R^2}{2} - \frac{r^2}{2}\right) = w_r$$

or

$$w_r = \frac{P_1 - P_2}{4\mu \ell}(R^2 - r^2) \qquad (5.24A)$$

The maximum velocity occurs at $r = 0$, and, therefore, may be written as

$$w_{max} = \frac{P_1 - P_2}{4\mu \ell} R^2 \qquad (5.24B)$$

Combining Equations (5.24A) and (5.24B), we obtain

$$w_r = w_{max} \frac{R^2 - r^2}{R} = w_{max}\left(1 - \frac{r^2}{R^2}\right) \qquad (5.25)$$

Equation (5.25) is known as *Stokes law for the parabolic velocity distribution* describing laminar flow in a pipe of constant cross section.

To evaluate the volumetric flow rate, examine Figure 5.2 in more detail. We will now consider only a small hollow cylinder of fluid having an inner radius r and outer radius $(r + dr)$. The cross-sectional area of the annular fluid ring is $dS = 2\pi r dr$ (examine both Figures 5.1 and 5.2). The volume rate passing through this section is

$$dV_{sec} = w_r dS = w_r 2\pi r dr \qquad (5.26)$$

Combining Equations (5.26) and (5.24A), the following expression is derived:

$$dV_{sec} = \frac{P_1 - P_2}{4\mu\ell}\int_0^R (R^2 - r^2)2\pi r dr$$

$$= \frac{P_1 - P_2}{4\mu\ell}\left(2\pi R^2 \int_0^R r dr - 2\pi \int_0^R r^3 dr\right) = \frac{P_1 - P_2}{8\mu\ell}\pi R^4 \qquad (5.27A)$$

Denoting pipe diameter $D = 2R$ and the pressure gradient $(p_1 - p_2) = \Delta P$, this expression can be rewritten as

$$V_{sec} = \frac{\pi D^4 \Delta P}{128\,\mu\ell} \qquad (5.27B)$$

Equations (5.27A) and (5.27B) express the volumetric rate for laminar flow in a straight circular pipe, and both are known as *Poiseuille's equation*.

The relationship between average velocity $\overline{W}$ and maximum velocity w_{max} may be obtained by comparing the following expressions:

$$V_{sec} = wS = w\pi R^2 \text{ and } \pi R^2 w = \frac{P_1 - P_2}{8\mu\ell}\pi R^4$$

Hence,

$$\overline{W} = \frac{P_1 - P_2}{8\mu\ell} R^2 \qquad (5.28)$$

Combining Equations (5.27B) and (5.28), we obtain

$$\overline{W} = \frac{w_{max}}{2} \quad (5.29)$$

Thus, the average velocity for laminar flow in a circular pipe is equal to half the maximum velocity at the pipe axis.

Similarly, the parabolic velocity distribution for pipe flow [Equation (5.25)] may be rewritten as follows:

$$w_r = 2\overline{W}\left(1 - \frac{r^2}{R^2}\right) \quad (5.30)$$

Further discussions are given by Knudsen and Katz (1958) and Curle and Davies (1968). Before proceeding, let us apply some of the above expressions to problem solving.

Example 5.3

Water at 86°F is flowing through a 7.62-cm i.d. pipe at a velocity of 0.3 m/s. Calculate the Reynolds number in both English and SI units.

Solution

The density and viscosity of water at 86°F (30°C) are as follows:
- density, $\varrho = 0.996 \ (62.43 \ \text{lb}_m/\text{ft}^3) = 62.18 \ \text{lb}_m/\text{ft}^3$
- viscosity, $\mu = 0.8007 \ \text{cp} \times (6.7197 \times 10^{-4} \ \text{lb}_m/\text{ft-s/cp}) = 5.38 \times 10^{-4} \ \text{lb}_m/\text{ft-s}$

The Reynolds number is defined by Equations (5.19):

$$Re = \frac{wD\varrho}{\mu}$$

Pipe diameter $D = 7.62 \ \text{cm} \times \text{in}/2.54 \ \text{cm} \times \text{ft}/12 \ \text{in} = 0.25 \ \text{ft}$. Velocity in pipe $\overline{W} = 0.3 \ \text{m/s} \times 3.28 \ \text{ft/m} = 0.984 \ \text{fps}$.

In English units,

$$Re = \frac{(0.984 \ \text{ft/s})(0.25 \ \text{ft})(62.18 \ \text{lb}_m/\text{ft}^3)}{5.38 \times 10^{-4} \ \text{lb}_m/\text{ft-s}} = 28,440$$

In SI units,

$$Re = \frac{(0.3 \ \text{m/s})(0.0762 \ \text{m})(996 \ \text{kg/m}^3)}{8.007 \times 10^{-4} \ \text{kg/m-s}} = 28,440$$

Hence, the flow is turbulent.

Example 5.4

Consider the system shown in Figure 5.3. A liquid is flowing through a 20-m long, 25.4-mm i.d. tube that is inclined 15° from a horizontal datum plane. The pressure at point 1 is measured to be 27 psia and the pressure at point 2 is 30 psia. The viscosity and density of the liquid are 0.07 N-s/m² and 996 kg/m³, respectively. Determine the following:

1 The direction of flow through the tube
2 The volumetric flow rate (assume the flow is laminar)

Solution

1. To determine the direction of flow, we must determine the energy at points 1 and 2.
At section 1,

$$P_1 = 27 \text{ psi} \times 6.985 = 186.2 \text{ kPa (or } 186{,}200 \text{ N/m}^2)$$
$$z = 20 \sin 15° = 5.18 \text{ m}$$
$$P_1 + \rho g z = 186{,}200 \text{ N/m}^2 + (996 \text{ kg/m}^3)(9.8 \text{ m/s}^2)(5.18 \text{ m})$$
$$= 236.8 \text{ kPa}$$

At section 2,

$$P_2 + \rho g z_0 = 30 \text{ psi} \times 6.985 + 0$$
$$= 206.9 \text{ kPa}$$

As the energy at point 1 is greater than at point 2, the flow is in the downward direction.

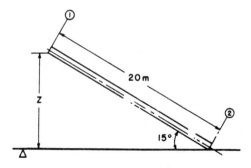

FIGURE 5.3. System under evaluation in Example 5.4.

2. To obtain the volumetric flow rate, we first compute the differential pressure gradient per unit length of tube:

$$\frac{d}{d\ell}(P + \gamma z) \frac{(236{,}800 - 206{,}900) \text{ N/m}^2}{20 \text{ m}} = 1495 \text{ N/m}^3$$

As the flow is laminar, we may use the Hagen-Poiseuille equation [Equation (5.27B)] to determine the volumetric flow rate:

$$V = \frac{\pi D_4}{128\,\mu} \frac{\Delta P}{\ell}$$

Pipe diameter $D = 63.5$ mm $= 0.0635$ m. Viscosity $\mu = 8.007 \times 10^{-4}$ kg/m-s. Hence

$$V = \frac{\pi(0.0254 \text{ m})^4}{128 \times 0.07 \text{ kg/m-s}} \left(1495 \frac{\text{kg}}{\text{m}^2\text{-s}^2}\right) = 2.182 \times 10^{-4} \text{ m}^3/\text{s}$$

As a check, we will compute the Reynolds number, $Re = D\overline{W}\varrho/\mu$.
Area of pipe

$$F = \frac{\pi}{4} D^2 = \frac{\pi}{4} (0.0254 \text{ m})^2 = 5.067 \times 10^{-4} \text{ m}^2$$

Velocity

$$\overline{W} = \frac{V}{F} = \frac{2.182 \times 10^{-4} \text{ m}^3/\text{s}}{5.067 \times 10^{-4} \text{ m}^2} = 0.431 \text{ m/s}$$

$$Re = \frac{(0.0254 \text{ m})(0.431 \text{ m/s})(996 \text{ kg/m}^3)}{0.07 \text{ kg/m-s}} = 156$$

The flow is indeed laminar. If the Reynolds number had exceeded 2000, the Hagen-Poiseuille equation would no longer apply.

Turbulence

The phenomenon of turbulence is important to many branches of chemical engineering. Despite its far-reaching importance, no theoretically rigorous expressions have been developed to describe it. Often, in the development of turbulent flow expressions, so many assumptions are made that it is difficult to establish whether agreement with experimental observation results from the

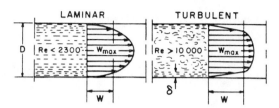

FIGURE 5.4. Velocity profiles for laminar and turbulent flows in a pipe.

application of reasonable simplifications or from the fortuitous cancellation of errors.

In turbulent flow, chaotic motion of fluid particles causes a smoothing of velocity streamlines in the bulk flow. In pipe flow, this results in a velocity distribution that is very different from the parabolic profile obtained with laminar flows. A comparison of laminar and turbulent velocity profiles for pipe flow is shown in Figure 5.4. Note that the turbulent distribution does not show any significantly wider apex. Experiments have demonstrated that unlike laminar flow, the average velocity is not equal to half the maximum. In fact, the average velocity is observed to be considerably greater and is a function of the Reynolds number, i.e., $w/w_{max} = f(Re)$. As a general rule, for turbulent Reynolds numbers up to 10,000, velocity $w \approx 0.8\ w_{max}$ and, at $Re \geq 10^8$, $w \approx 0.9\ w_{max}$.

Present understanding of turbulence does not enable velocity profiles to be computed a priori, as with laminar flow. The turbulent profile illustrated in Figure 5.4 does not represent an instantaneous distribution but rather one that is time-averaged. The instantaneous velocity at a point varies with time in both magnitude and direction. Therefore, it is an irregularly oscillating function that is best illustrated by Figure 5.5. However, we can define a time-smoothed velocity $\overline{W}_2$ in the x-direction by taking a time-average of w_x over some time interval t. The time interval must be large with respect to the time of turbulent oscillation but small with respect to actual time changes:

$$\overline{W}_x = \frac{\int_0^\tau w_x dt}{t} \qquad (5.31)$$

where w_x is the instantaneous velocity (a function of time t). w_x is the sum of the time-smoothed velocity $\overline{W}_x$ and a velocity fluctuation Δw:

$$w_x = \overline{W}_x \pm \Delta w \qquad (5.32)$$

The value of the time-averaged velocity $\overline{W}_x$ over a sufficiently large time

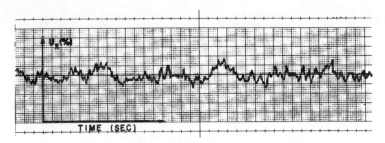

(a) OSCILLOGRAPH PRINTOUT FROM HOT WIRE ANEMOMETER SHOWING VELOCITY FLUCTUATIONS.

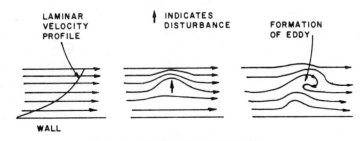

(b) DISTURBANCE INTRODUCED TO FLUID CAUSING STREAMLINE INTERMIXING & EDDY FORMATION.

FIGURE 5.5. Actual (instantaneous) and time-averaged velocities for turbulent flow.

interval is constant. Since the frequency of the velocity fluctuation is extremely large, the time interval needs to be only a few seconds or even fractions of seconds. Consequently, we may consider a time-independent change of the velocities averaged over the cross section of piping, rather than the change of instantaneous values. That is, we shall consider local flow to be at quasi-stationary motion. Although the instantaneous fluctuations are small, they have a dramatic effect on the flow characteristics. For example, the transverse movement of a fluid particle from a faster-moving region to a slower-moving one has the effect of increasing the velocity in the slower region, thus acting as an equivalent shear stress. This turbulent shear stress may be hundreds of times greater than the laminar stress due to the sliding of one fluid layer over another.

To evaluate the importance of fluctuating velocities relative to their average, turbulence is often characterized by a parameter known as the *intensity of turbulence* (or intensity factor):

$$I_T = \frac{\overline{\Delta W}}{\overline{W}} \tag{5.33}$$

where $\overline{\Delta W}$ is the root-average-square of fluctuating velocities, expressed as

$$I_T = \frac{\sqrt{\frac{1}{3}(\overline{\Delta W_x^2} + \overline{\Delta W_y^2} + \overline{\Delta W_z^2})}}{\overline{W}_x} \tag{5.34}$$

The intensity of turbulence *is an estimate of the fluctuations at a given point within the flow*. For turbulent flow in piping, $I_T \approx 0.01{-}0.1$.

The special case in which the three average-square fluctuating velocities are equal is referred to as *isotropic turbulence*. For this case, Equation (5.34) simplifies to

$$I_T = \sqrt{\overline{\Delta W_x^2}} \tag{5.35}$$

In a practical sense, turbulent flows almost never reach isotropic conditions. In pipe flow, for example, the isotropic state is only approached at the axis of flow. Greater deviation from isotropic turbulence occurs in fluid regions removed from the axis, particularly near the wall. The intensity factor alone does not completely describe turbulent motion. Methods are available for specifying the turbulence scale and turbulent viscosity.

The closer two fluid particles are to each other, the closer their actual or instantaneous velocities. In contrast, no relation exists between the velocities of two fluid particles relatively removed from one another. Hence, we can assume that the fluid particles belong to different groups of fluid elements, which tend to move together and form some unified aggregates. These aggregates are called *eddies*. The size of eddies or the depth of their penetration before collapsing (which may be identified approximately by the distance between two close particles not belonging to the same eddy) depends on the rate of turbulence. This rate of turbulence is known as the *scale*. A correlation coefficient is conveniently defined as follows:

$$R(y) = \frac{\overline{\Delta W_{x_1} \Delta W_{x_2}}}{\sqrt{\overline{\Delta W_{x_1}^2}} \sqrt{\overline{\Delta W_{x_2}^2}}} \tag{5.36}$$

For flow in the *x*-direction, $\overline{\Delta W}_{x_1}$ and $\overline{\Delta W}_{x_2}$ are the fluctuating velocities occurring at the same time t_1, at points 1 and 2, separated by distance y.

The turbulence scale is based on the correlation coefficient $R(y)$, expressed as a function between two points:

$$L_T = \int_0^\infty R(y)dy \qquad (5.37)$$

Note that the term "eddy" is not inclusive of turbulence. Eddy motion or currents may also exist in laminar flow because they are characterized simply by velocity differences over the flow area. The distinction between turbulent and laminar flows, therefore, is not denoted by eddy motion, but by the presence of chaotic fluctuations of velocities at different points throughout the flow area. These fluctuations produce particle displacements in a direction normal to the axis of flow.

To describe these chaotic fluctuations, we need a parameter that relates the viscous forces of the fluid to its kinetic energy. In turbulent flow, this parameter is the eddy viscosity. Let us consider two fluid particles flowing in the x-direction parallel to the pipe axis. Earlier, we defined y as the distance between these particles in the direction normal to the pipe axis. The particles' velocity components in the flow direction are $\overline{W}_{x_1}$ and $\overline{W}_{x_2}$, which differ from each other by $d\overline{W}_x$. Because of this difference, a shear stress results:

$$\tau_N = -\mu \frac{d\overline{W}_x}{dy} = -\varrho \nu \frac{d\overline{W}_x}{dy} \qquad (5.38)$$

where μ and ν are the dynamic and kinematic viscosities, respectively. ϱ is density, and subscript "N" denotes that we are dealing with a Newtonian fluid. Equation (5.38) is a general expression that applies only to laminar flow. In laminar flow, τ_N is the only stress existing between fluid layers located distance dy apart. In turbulent flow, however, fluid particles move relative to each other, not only in the longitudinal direction (i.e., together with the flow), but also in the transverse direction. This creates an additional shear stress τ_T (subscript "T" refers to turbulent), which, by analogy with τ_N, may be expressed as follows:

$$\tau_T = -E_\nu \frac{d\overline{W}_x}{dy} \qquad (5.39)$$

Parameter E_ν in Equation (5.39) is called the *eddy viscosity* and is analogous to dynamic viscosity μ. The quantity $\epsilon_m = E_\nu/\varrho$ is the *eddy diffusivity of momentum,* which is analogous to kinematic viscosity. These two fluid properties (E_ν and ϵ_m) depend on the fluid velocity and geometry of the flow system. That is, they are functions of all factors that influence the nature of turbulence and the deviation of velocities. In particular, they are sensitive to location within the turbulent field, i.e., distance from the wall, etc. Thus, the total shear stress in a turbulent flow is the sum of viscous stresses and turbulent stresses:

$$\tau = \tau_N + \tau_T = -\varrho(\nu + \nu_T)\frac{d\overline{W}_x}{dy} \quad (5.40)$$

Examining the turbulent velocity profile again in Figure 5.4, we note that the time-averaged velocities comprise a relatively flat distribution in the bulk flow. It is only in the vicinity of the wall that velocities decrease rapidly and eventually become zero at the solid boundary. In the immediate vicinity of the wall, the fluid motion becomes less turbulent and more laminar. The solid boundary has the effect of dampening turbulent fluctuations in the transverse direction. Thus, turbulent motion never really exists in pure form; rather, it is accompanied by laminar motion.

Fluid motion is divided conditionally into two zones: the central or *bulk flow,* where the motion is turbulent, and the *hydrodynamic boundary layer* which is the zone near the wall where turbulent flow changes into laminar flow. Within the hydrodynamic boundary layer, a thin sublayer exists very close to the pipe wall. In this region, viscous forces have a predominant influence on the fluid motion. The velocity gradient in the *laminar boundary sublayer* is very high. In turbulent flow the laminar sublayer is very thin—sometimes fractions of a millimeter in thickness—and decreases with increasing turbulence. Regardless of the thickness, the sublayer has a dramatic influence not only on the hydraulic resistance of fluid motion, but on the mechanisms of heat and mass transfer.

Between the bulk flow and the laminar sublayer there exists a region known as the buffer zone. The laminar sublayer and buffer zone make up the hydrodynamic boundary layer. The thickness of this layer is determined by shear stresses that exist between fluid particles caused by viscous forces and turbulent fluctuations. Hence, in this region the magnitudes of ν and ν_T become comparable. Further discussions are given by Knudsen and Katz (1958), Rohsenow and Hartnett (1973), and Perry (1950).

CONTINUITY EQUATION

We shall develop the general relationship among the velocities of flow through any system. Following the analysis presented in the previous chapter, Figure 5.6 shows a differential fluid element of volume $dV = dxdydz$. The

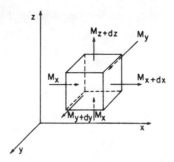

FIGURE 5.6. Fluid element showing mass transfer through defined boundaries.

velocity component along the x-axis is w_x. The input per unit time dt in the x-direction through the left face of the element ($dS = dydz$) is equal to

$$M_x = \varrho w_x dydzdt \quad (5.41)$$

where ϱ = fluid density at the left face of the element.

The velocity and density at the parallel face (located at $x + dx$) are $[w_x + \partial(\varrho w_x)dx/\partial x]$ and $(\varrho + \partial \varrho dx/\partial x)$, respectively. Hence, the output at the right face of the element in the same time interval dt is

$$M_{x+dx} = \left[\varrho w_x + \frac{\partial(\varrho w_x)}{\partial x} dx\right] dydzdt \quad (5.42)$$

The mass differential in the element along the x-axis is

$$dM_x = M_x - M_{x+dx} = -\frac{\partial(\varrho w_x)}{\partial x} dxdydzdt \quad (5.43)$$

and the total mass differential in the elemental volume dV over time dt is

$$dM = -\left[\frac{\partial(\varrho w_x)}{\partial x} + \frac{\partial(\varrho w_y)}{\partial y} + \frac{\partial(\varrho w_z)}{\partial z}\right] dxdydzdt \quad (5.44)$$

For a mass differential to occur in the fluid element, density must also change with respect to time. Hence,

$$dM = \frac{\partial \varrho}{\partial t} dxdydzdt \quad (5.45)$$

Equating these last two expressions through *dM* and rearranging terms, we obtain

$$\frac{\partial \varrho}{\partial t} + \frac{\partial(\varrho w_x)}{\partial x} + \frac{\partial(\varrho w_y)}{\partial y} + \frac{\partial(\varrho w_z)}{\partial z} = 0 \qquad (5.46)$$

Equation (5.46) is the differential equation of continuity for unsteady flow of a compressible fluid. The equation may be written in another form by differentiating the product ϱw:

$$\frac{\partial \varrho}{\partial t} + \frac{\partial \varrho}{\partial x} w_x + \frac{\partial \varrho}{\partial y} w_y + \frac{\partial \varrho}{\partial z} w_z + \frac{\partial w_x}{\partial x} \varrho + \frac{\partial w_y}{\partial y} \varrho + \frac{\partial w_z}{\partial z} \varrho = 0 \qquad (5.47A)$$

or

$$\frac{1}{\varrho}\frac{d\varrho}{dt} + \frac{\partial w_x}{\partial x} + \frac{\partial w_y}{\partial t} + \frac{\partial w_z}{\partial z} = 0 \qquad (5.47B)$$

where $\partial \varrho / \partial t$ is a substantial derivative. In steady-state flows, density ϱ does not change with time, i.e., $\partial \varrho / \partial t = 0$, and Equation (5.46) simplifies to the following:

$$\frac{\partial(\varrho w_x)}{\partial x} + \frac{\partial(\varrho w_y)}{\partial y} + \frac{\partial(\varrho w_z)}{\partial z} = 0 \qquad (5.48)$$

For liquids, which are practically incompressible, and for gases under isothermal conditions and at velocities much less than velocity of sound, ϱ = constant. Hence, Equation (5.48) is simply

$$\frac{\partial w_x}{\partial x} + \frac{\partial w_y}{\partial y} + \frac{\partial w_z}{\partial z} = 0 \qquad (5.49)$$

Equation (5.49) is the differential continuity equation for the flow of incompressible fluids.

EQUATIONS OF MOTION FOR IDEAL FLUIDS

Following the analysis of Streeter and Wylie (1979), we shall consider steady flow of an ideal fluid having zero viscosity. As shown in Chapter 4, the projections of pressure and gravity forces onto the coordinate axes acting on the elemental volume $dV = dxdydz$ are the following:

For the x-axis,

$$-\frac{\partial P}{\partial x} dxdydz$$

For the y-axis,

$$-\frac{\partial P}{\partial y} dxdydz$$

For the z-axis,

$$-\left(\varrho g + \frac{\partial P}{\partial z}\right) dxdydz$$

In accordance with the basic principle of dynamics, the sum of the projection of forces acting on the elemental volume in motion is equal to the product of mass and acceleration. Hence, for the x, y, and z axis projections, we may write

$$\left.\begin{array}{l} \varrho dxdydz \dfrac{dw_x}{dt} = -\dfrac{\partial P}{\partial x} dxdydz \\[1em] \varrho dxdydz \dfrac{dw_y}{dt} = -\dfrac{\partial P}{\partial x} dxdydz \\[1em] \varrho dxdydz \dfrac{dw_y}{dt} = -\dfrac{\partial P}{\partial x} dxdydz \end{array}\right\} \quad (5.50)$$

And, after simplifying,

$$\left.\begin{array}{l} \varrho \dfrac{dw_x}{dt} = -\dfrac{\partial P}{\partial x} \\[1em] \varrho \dfrac{dw_y}{dt} = -\dfrac{\partial P}{\partial y} \\[1em] \varrho \dfrac{dw_z}{dt} = -\varrho g - \dfrac{\partial P}{\partial z} \end{array}\right\} \quad (5.51)$$

The substantial derivatives of the velocity components are

$$\left.\begin{array}{l} \dfrac{dw_x}{dt} = \dfrac{\partial w_x}{\partial x} w_x + \dfrac{\partial w_x}{\partial y} w_y + \dfrac{\partial w_x}{\partial z} w_z \\[6pt] \dfrac{dw_y}{dt} = \dfrac{\partial w_y}{\partial x} w_x + \dfrac{\partial w_y}{\partial y} w_y + \dfrac{\partial w_y}{\partial z} w_z \\[6pt] \dfrac{dw_z}{dt} = \dfrac{\partial w_z}{\partial x} w_x + \dfrac{\partial w_z}{\partial y} w_y + \dfrac{\partial w_z}{\partial z} w_z \end{array}\right\} \quad (5.52)$$

The above systems of Equations (5.51) and (5.52) are known as Euler's differential equations of motion for an ideal fluid under steady flow. In unsteady flow, velocity changes not only when fluid particles move from one point in space to another, but also with time. Therefore, the components of acceleration in Equation (5.51) must be expressed in the following form:

$$\left.\begin{array}{l} \dfrac{dw_x}{dt} = \dfrac{\partial w_x}{\partial t} + \dfrac{\partial w_x}{\partial x} w_x + \dfrac{\partial w_x}{\partial y} w_y + \dfrac{\partial w_x}{\partial z} w_x \\[6pt] \dfrac{dw_y}{dt} = \dfrac{\partial w_y}{\partial t} + \dfrac{\partial w_y}{\partial x} w_x + \dfrac{\partial w_y}{\partial y} w_y + \dfrac{\partial w_y}{\partial z} w_x \\[6pt] \dfrac{dw_z}{dt} = \dfrac{\partial w_z}{\partial t} + \dfrac{\partial w_z}{\partial x} w_x + \dfrac{\partial w_z}{\partial y} w_y + \dfrac{\partial w_z}{\partial z} w_x \end{array}\right\} \quad (5.53)$$

The systems of Equations (5.51) and (5.53) represent Euler's differential equations of motion for unsteady flow.

DIFFERENTIAL EQUATIONS FOR VISCOUS FLUIDS

Navier-Stokes Equations

The motion of real or viscous fluids is influenced by frictional forces in addition to pressure and gravity. Frictional forces τ, acting on a volume element $dV = dxdydz$ shown in Figure 5.7, result in shear stresses τ forming on the faces. For simplicity, first consider one-dimensional flow in the x-direction where the velocity projection w_x depends only on distance z above some horizontal reference plane. In this case, the shear stresses on the lower and upper faces of the volume element $dF = dxdy$ are τ and $\tau + (\partial \tau / \partial z) dz$, respectively. The partial derivative $\partial \tau / \partial z$ expresses the change in shear stress along

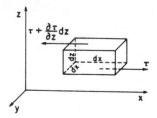

FIGURE 5.7. Forces acting on a volume element of moving fluid.

the z-axis at points located on the lower face of the volume element, and $(\partial\tau/\partial z)dz$ is the change of shear stress along the total length of the element's edge dz. The projection of the resultant shear forces in the x-direction is

$$\tau dxdy - \left(\tau + \frac{\partial \tau}{\partial z} dz\right) dxdy = -\frac{\partial \tau}{\partial z} dxdydz \qquad (5.54)$$

From Newton's law $\tau = -\mu\ \partial w_x/\partial z$, we may rewrite this expression as follows:

$$\mu \frac{\partial \left(\frac{\partial w_x}{\partial z}\right)}{\partial z} dxdydz = \mu \frac{\partial^2 w_x}{\partial z^2} dxdydz \qquad (5.55)$$

For three-dimensional flow, component w_x changes along three coordinate axes, and the projection of the resultant shear forces in the x-direction is

$$\mu \left(\frac{\partial^2 w_x}{\partial z^2} + \frac{\partial^2 w_x}{\partial y^2} + \frac{\partial^2 w_x}{\partial x^2}\right) dxdydz$$

or

$$\frac{\partial^2 w_x}{\partial x^2} + \frac{\partial^2 w_x}{\partial y^2} + \frac{\partial^2 w_x}{\partial z^2} = \nabla^2 w_x$$

where ∇^2 is called the Laplacian operator and is defined as

$$\nabla^2 = \frac{\partial^2}{\partial x^2} + \frac{\partial^2}{\partial y^2} + \frac{\partial^2}{\partial z^2} \qquad (5.56)$$

Using this notation, expressions can be written for the resultant projections of gravity, pressure, and shear forces on respective coordinate axes:

For the x-axis,
$$\left(-\frac{\partial P}{\partial x} + \mu \nabla^2 w_x\right) dxdydz$$

For the y-axis,
$$\left(-\frac{\partial P}{\partial x} + \mu \nabla^2 w_y\right) dxdydz$$

For the z-axis,
$$\left(-\varrho g - \frac{\partial P}{\partial x} + \mu \nabla^2 w_x\right) dxdydz$$

Note that the gravity and pressure force projections are taken from Euler's equations.

From dynamics, the sum of the projection of forces equals the product of the fluid mass $\varrho dxdydz$ in the elemental volume and the projections of acceleration. Applying this principle and after simplification, we obtain the celebrated Navier-Stokes equations of motion for viscous incompressible fluids expressed in Cartesian coordinates:

$$\left. \begin{array}{l} \varrho \dfrac{dw_x}{d\tau} = -\dfrac{\partial P}{\partial x} + \mu \nabla^2 w_x \\[2mm] \varrho \dfrac{dw_y}{d\tau} = -\dfrac{\partial P}{\partial y} + \mu \nabla^2 w_y \\[2mm] \varrho \dfrac{dw_z}{d\tau} = -\varrho g - \dfrac{\partial P}{\partial z} + \mu \nabla^2 w_z \end{array} \right\} \quad (5.57)$$

The Navier-Stokes equations for compressible fluids are as follows:

$$\left. \begin{array}{l} \varrho \dfrac{dw_x}{d\tau} = -\dfrac{\partial P}{\partial x} + \mu \left(\nabla^2 w_x + \dfrac{1}{3}\dfrac{\partial \theta}{\partial x}\right) \\[2mm] \varrho \dfrac{dw_y}{d\tau} = -\dfrac{\partial P}{\partial y} + \mu \left(\nabla^2 w_y + \dfrac{1}{3}\dfrac{\partial \theta}{\partial y}\right) \\[2mm] \varrho \dfrac{dw_z}{d\tau} = -\varrho g - \dfrac{\partial P}{\partial z} + \mu \left(\nabla^2 w_z + \dfrac{1}{3}\dfrac{\partial \theta}{\partial x}\right) \end{array} \right\} \quad (5.58)$$

The partial derivatives $\partial\theta/\partial x$, $\partial\theta/\partial y$, and $\partial\theta/\partial z$ represent changes in velocities along the x-, y-, and z-axes and are associated with compression and tension forces acting on the fluid.

The terms on the left-hand sides (LHS) of Equations (5.57) and (5.58) represent the rate of inertial forces for the moving fluid. Terms on the RHS describe three types of forces:

1. ϱg denotes gravitational forces acting on the fluid volume element.
2. The partial derivatives $\partial P/\partial x$, $\partial P/\partial y$ and $\partial P/\partial z$ denote pressure forces.
3. The product of viscosity and the sum of the second derivatives of the velocity projections denote the viscous forces acting on the moving fluid element.

All terms in these two sets of equations have dimensions of force per unit volume. For ideal fluids, i.e., for $\mu = 0$, where there are no viscous forces, our expressions reduce to Euler's equations [Equations (5.52) and (5.53)].

The Navier-Stokes equations represent a system of nonlinear partial expressions having no general solution. In fact, these equations are often of a higher order because viscosity can be a function of temperature, as well as velocity, for many flow systems. Two general approaches may be used in applying the Navier-Stokes equations to analyzing flow systems. The expressions may be reduced to specific solutions either by application of a series of simplifying assumptions or by transformation techniques based on the theory of similarity. The first approach is illustrated in the following example.

Example 5.5

Derive an expression for pressure drop for steady-state viscous flow in a horizontal tube of radius r. The flow is in one direction and is driven only by a constant-pressure differential. The derivation should be based on an incompressible fluid with constant viscosity.

Solution

Any model or expression developed from first principles will only have validity within the limitations of the constraints imposed and the assumptions applied to its development. It is important that these assumptions be stated clearly along with the final expression to prevent improper application and, thus, erroneous results.

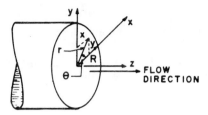

FIGURE 5.8. Flow system under evaluation in Example 5.5.

The constraints applied to our derivation are:

1 Constant viscosity
2 Incompressibility
3 One-directional flow
4 Steady-state flow

For simplicity, we shall apply the assumption that the fluid is far from the tube's inlet and exit; hence, end effects will be ignored.

With the constraints and major assumptions established, we now must define a convenient coordinate system. As the flow is specified as being in one direction, the z-axis is assigned as the axis of symmetry of flow. Also, y shall denote the vertical direction and x the horizontal coordinate. The system under evaluation is represented by the sketch in Figure 5.8.

As w_x and w_y are zero (constraint 3), the continuity equation reduces to

$$\frac{\partial w_z}{\partial z} = 0$$

and, for steady-state flow (constraint 4),

$$\frac{\partial w_z}{\partial t} = 0$$

Substituting these terms and assumptions into Equation (5.57) and noting that the flow is horizontal, we obtain the following:

$$\frac{dP}{dz} = \mu \left(\frac{\partial^2 w_z}{\partial x^2} + \frac{\partial^2 w_z}{\partial y^2} \right)$$

Because of the geometry of this flow system, it is convenient to transform our starting expression into cylindrical coordinates:

$$z = z \quad y = r\sin\theta \quad \theta = \tan^{-1} y/x$$

$$x = r\cos\theta \quad r = \sqrt{x^2 + y^2}$$

Substituting these into our starting expression,

$$\frac{1}{\mu}\frac{dP}{dz} = \frac{\partial^2 w_z}{\partial r^2} + \frac{1}{r}\frac{\partial w_z}{\partial r} + \frac{1}{r^2}\frac{\partial^2 w_z}{\partial \theta^2}$$

Because the z-axis has been assigned as the axis of symmetry of the flow,

$$\frac{\partial^2 w_z}{\partial \theta^2} = 0$$

As the flow is based on a constant pressure differential $1/\mu\, dP/dz = $ constant C:

$$C = \frac{d^2 w_r}{dr^2} + \frac{1}{r}\frac{dw_r}{dr}$$

$$= \frac{1}{r}\frac{d}{dr}\left(r\frac{dw_r}{dr}\right)$$

The above expression can now be integrated twice to obtain an expression for the velocity profile. For the first integration, apply the boundary condition:

$$\frac{dw_z}{dr} = 0 \text{ at } r = 0$$

For the second integration, apply the boundary condition:

$$w_z = 0 \text{ at } r = R$$

where R is the tube radius. The resultant velocity profile is

$$w_z = \frac{1}{4\mu}\frac{dP}{dz}(r^2 - R^2)$$

This is the same as Equation (5.25A). Following the same development, we can also obtain an expression for the maximum velocity. w_{zmax} occurs at

$r = 0$, see Equation (5.24B). The above expression may be integrated over the tube cross section to obtain the average velocity:

$$\overline{W}_z = - \frac{R^2}{8\mu} \frac{dP}{dz}$$

And, finally, we can integrate this last expression over the limits of $z = 0$ for $p = p_1$ to $z = L$ for $p = p_2$, to obtain pressure drop:

$$p_1 - p_2 = \frac{8\mu \overline{W}_z L}{R^2}$$

This expression is the Hagen-Poiseuille equation derived earlier and, of course, is applicable only to laminar flow.

Additional illustrative examples are given by Bird et al. (1960), Curle and Davies (1968), Brodkey (1967), and Cheremisinoff (1979, 1981a,b). Problems at the end of this chapter provide additional instruction.

We will now use principles of similarity theory to transform the Navier-Stokes equations into forms suitable for flow analysis.

Transformation Techniques

Following the methods outlined by Sedov (1959) and Gukhman (1965), we will make certain transformations to the system of Equations (5.58), making use of the following characteristic parameters as measured units of the appropriate flow variables: ϱ_0, P_0, μ_0, τ_0, w_0, and ℓ_0. This enables us to define the following scaling factors:

- $\varrho/\varrho_0 = P'$
- $P/P_0 = p$
- $\mu/\mu_0 = M'$
- $\tau/\tau_0 = T'$
- $w/w_0 = \tilde{W}$
- $w_x/w_{x0} = W_x$
- $w_y/w_{y0} = W_y$
- $w_z/w_{z0} = W_z$
- $\ell/\ell_0 = x/x_0 = X'$
- $y/y_0 = Y'$
- $z/z_0 = Z'$

By rearranging Equations (5.58),

$$\varrho\left[\frac{\partial w_x}{\partial t} + w_x\frac{\partial w_x}{\partial x} + w_y\frac{\partial w_y}{\partial y} + w_z\frac{\partial w_z}{\partial z}\right]$$
$$= \varrho g - \frac{\partial P}{\partial x} + \mu\left(\nabla^2 w_x + \frac{1}{3}\frac{\partial\theta}{\partial x}\right) \qquad (5.59)$$

Then, substituting the above scaling factors for each variable, we obtain the following:

$$P\varrho_0\left[\frac{\partial W_x}{\partial T'}\cdot\frac{w_0}{T_0} + W_x\frac{\partial W_x}{\partial x}\cdot\frac{w_0}{\ell_0} + w_0 W_y\frac{\partial W_y}{\partial y}\cdot\frac{w_0}{\ell_0} + w_0 W_z\frac{\partial W_z}{\partial z}\cdot\frac{w_0}{\ell_0}\right]$$
$$= P'\varrho_0 g - \frac{\partial p}{\partial x}\frac{P_0}{\ell_0} + \mu M'\left(\nabla^2 W_x\frac{w_0}{\ell_0^2} + \frac{1}{3}\frac{\partial\theta}{\partial x}\cdot\frac{w_0}{\ell_0^2}\right) \qquad (5.60)$$

where

$$\theta = \frac{\partial W_x}{\partial x} + \frac{\partial W_y}{\partial y} + \frac{\partial W_z}{\partial z} = \text{div } W$$

and

$$\nabla^2 W_x = \frac{\partial^2 W_x}{\partial x^2} + \frac{\partial^2 W_y}{\partial y^2} + \frac{\partial^2 W_z}{\partial z^2}$$

As ϱ_0, μ_1, and ℓ_0 are constants, they may be combined into a single coefficient:

$$\frac{\varrho_0 W_0^2}{\ell_0}$$

To do this, w_0^2/ℓ_0 must be taken outside of the brackets on the LHS, while w_0/ℓ_0^2 is removed to the outside of the brackets on the last term in the RHS of Equations (5.60). Our expression then becomes

$$\left(\frac{\varrho_0 w_0^2}{\ell_0}\right)P'\left[\frac{\partial W_x}{\partial T'}\left(\frac{\ell_0}{w_0 T_0}\right) + W_x\frac{\partial W_x}{\partial x} + W_y\frac{\partial W_x}{\partial y} + W_z\frac{\partial W_z}{\partial z}\right]$$
$$= (\varrho_0 g)P' - \left(\frac{P_0}{\ell_0}\right)\frac{\partial p}{\partial x} + \frac{\mu_0 w_0}{\ell_0^2}M'\left[\nabla^2 W_x + \frac{1}{3}\frac{\partial\theta}{\partial x}\right] \qquad (5.61)$$

Dividing both sides of the equation by $\varrho_0 w_0^2/\ell_0$, we obtain

$$P\left[\frac{\partial W_x}{\partial T'}\left(\frac{\ell_0}{w_0 T_0}\right) + W_x\frac{\partial W_x}{\partial x} + W_y\frac{\partial W_x}{\partial y} + W_z\frac{\partial W_x}{\partial z}\right]$$
$$= \frac{g_0 \ell_0}{w_0^2} P' - \left(\frac{P_0}{\varrho_0 w_0^2}\right)\frac{\partial p}{\partial x} + \left(\frac{\mu_0}{\ell_0 w_0 \varrho_0}\right) M'\left[\nabla^2 W_x + \frac{1}{3}\frac{\partial \theta}{\partial x}\right]$$
(5.62A)

Equation (5.62A) contains only similarity invariants (i.e., characteristic parameters), simplexes and complexes. These characteristic parameters now may be expressed in terms of recognizable dimensionless groups.

The following dimensionless groups are recognized in Equations (5.62A):

$$\frac{w\tau}{\ell} = H_o = \text{Homochronicity number}$$

$$\frac{w^2}{g\ell} = Fr = \text{Froude number}$$

$$\frac{\ell w \varrho}{\mu} = Re = \text{Reynolds Number}$$

$$\frac{P}{\varrho w^2} = Eu = \text{Euler number}$$

Thererfore, we may rewrite Equation (5.62A) in the notation of identified dimensionless groups:

$$P\left[\frac{\partial w_x}{\partial T}\cdot\frac{1}{H_o} + w_x\frac{\partial w_x}{\partial x} + w_y\frac{\partial w_y}{\partial y} + w_z\frac{\partial w_z}{\partial z}\right]$$
$$= \frac{1}{Fr_o} P - Eu_o \frac{\partial p}{\partial x} + \frac{1}{Re_o} M'\left[\nabla^2 W_x + \frac{1}{3}\frac{\partial \theta}{\partial x}\right]$$
(5.62B)

Hence, in applying the Navier-Stokes equations to analyze a flow system, the equations may be reduced to a workable solution by properly defining the physics of the system and, consequently, neglecting the least important dimensionless groups. For example, if gravity forces are unimportant (as in Example 5.5), then the Froude number may be neglected and the conditions of identity of a system of Navier-Stokes equations are determined strictly by the Homochronicity, Reynolds, and Euler numbers.

For dynamically possible flows at the condition of similarity of uniqueness, the identification of the Homochronicity and Reynolds numbers is necessary to the existence of kinematic and dynamic similarity. It follows that the Euler number is not a premise but a consequence of similarity existence. Therefore, the Euler number should be considered as a nondetermining dimensionless group. Dimensionless groups of this type are unique functions of other determining groups. For this reason, the Navier-Stokes equations may be represented in the following dimensionless form:

$$Eu = \phi(H_o, Fr, Re) \qquad (5.63)$$

As noted earlier, the equations of motion alone are insufficient to describe a flow system. A second equation needed to characterize a flow system is the continuity equation [Equation (5.49)]. Continuity is dependent only on the Homochronicity number:

$$\frac{1}{H_o}\frac{\partial P}{\partial T'} + \frac{\partial(\varrho w_x)}{\partial x} + \frac{\partial(\varrho w_y)}{\partial y} + \frac{\partial(\varrho w_z)}{\partial z} = 0 \qquad (5.64)$$

At steady-state, H_o is eliminated and similarity in the absence of viscous forces (i.e., the Froude number) is provided by identifying the Reynolds number. The Euler number, in this case, is a unique function of the Reynolds number, whence Equation (5.63) reduces to

$$Eu = \phi(Re) \qquad (5.65)$$

This last statement reflects only the kinematic and dynamic similarities of the flow. For total similarity it is also necessary to obtain geometric similarity. For flow in round pipes, the simplex of geometric similarity is the ratio of ℓ/d. Denoting this simplex number as $\tilde{\Gamma} = \ell/d$, it is included in Equation (5.65) to give the following:

$$Eu = \phi(Re, \tilde{\Gamma}) \qquad (5.66)$$

THE TOTAL ENERGY BALANCE

Often the equations of motion and continuity are insufficient to define a flow system. Another equation is usually required to solve for the flow system parameters of interest. For many flow systems, this third equation is the energy equation. Consider a steady continuous-flow system comprising a pump and a heat exchanger, as shown in Figure 5.9. Because there is no

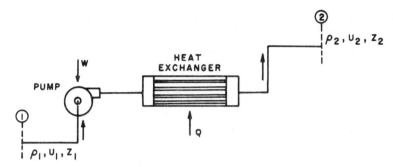

FIGURE 5.9. Flow system consisting of pump and heat exchanger.

accumulation of energy in this system, the overall energy balance is developed in the following manner.

The energy input to the system is the sum of the kinetic E_{k_1}, potential E_{p_1}, volumetric E_{v_1}, and internal E_{i_1} energies at section 1; the heat $\dot{Q}$, added through the exchanger; and the mechanical work W', performed on the fluid by the pump. The energy output comprises kinetic E_{k_2}, potential E_{p_2}, volumetric E_{v_2}, and internal E_{i_2} energies at section 2. Hence, we may equate the energy input to the system output to develop the energy balance for Figure 5.9.

$$E_{k_1} + E_{p_1} + E_{v_1} + E_1 + \dot{Q} + W' = E_{k_2} + E_{p_2} + E_{v_2} + E_2 \quad (5.67)$$

For a basis of calculation, we will apply Equation (5.67) to a fluid mass of 1 kg of weight. (Note in the gravitational unit system that a kilogram is a unit of force.) Now consider each of the energy components in Equation (5.67).

Potential energy is the product of weight (1 kg) and the fluid's elevation above some specified datum plane. That is,

$$E_{p_1} = Z_1; \quad E_{p_2} = Z_2 \quad (5.68)$$

where Z_1 and Z_2 are elevations at sections 1 and 2 relative to some reference plane.

The volumetric energy under pressure p is equivalent to the work expended to form volume v' at this pressure. The volumetric energies of 1 kg of fluid at the two sections are

$$E_{v_1} = P_1 v'_1; \quad E_{v_2} + P_2 v'_2 \quad (5.69)$$

Kinetic energy is the product of mass and one-half the square of the linear velocity of the fluid. In this case, the weight of the fluid, i.e., the product of fluid mass and its acceleration due to gravity ($g = 9.81$ m/s²), is equal to one kilogram-force. The mass of the liquid expressed in gravitational units is $1/g \times kg \times s^2/m^4$. If the fluid flows with velocity w, then its kinetic energy is $w^2/2g$ and, for sections 1 and 2, we obtain

$$E_{k_1} = \frac{\overline{w}_1^2}{2g\alpha_1} \; ; \; E_{k_2} = \frac{\overline{w}_2^2}{2g\alpha_2} \qquad (5.70)$$

α is a correction coefficient for inaccuracies in measuring the average velocities. For turbulent flows, $\alpha = 1$.

The fourth term in Equation (5.67) is the internal energy, which is a thermodynamic property of the system. It is defined relative to some specified reference state, usually at $t = 0°C$ and $p = 1$ atm. It is convenient to introduce a value that expresses the change of the internal energy of 1 kg of fluid as it passes through the system $(E_2 - E_1)$.

Denoting W' as the work performed in pumping 1 kg of fluid and $\dot{Q}$ as the heat added in units of work, Equation (5.67) may be rewritten as follows:

$$P_1 v_1' + \frac{\overline{w}_1^2}{2g\alpha_1} + Z_1 + \dot{Q} + W' = P_2 v_2' +$$
$$+ \frac{\overline{w}_2^2}{2g\alpha_2} + Z_2 + (E_2 - E_1) \qquad (5.71)$$

Equation (5.71) is known as the law of conservation of energy and may be applied to both compressible and incompressible flows. Each term has units of length m or, more precisely, units of energy per unit weight (kg × m/kg = m).

For engineering calculations, the internal energy terms may be approximated by enthalpy. For this system, enthalpy must be expressed as follows:

$$i_1 = E_1 + P_1 v_1'; \; i_2 = E_2 + P_2 v_2' \qquad (5.72)$$

Substituting Equation (5.72) into Equation (5.71), we obtain

$$\frac{\overline{w}_1^2}{2g\alpha_1} + Z_1 + \dot{Q} + W' = \frac{\overline{w}_2^2}{2g\alpha_1} + Z_2 + (i_2 - i_1) \qquad (5.73)$$

For ideal gases, the change in enthalpy may be determined from the product of heat capacity at constant pressure and the system's temperature differential, $C_p(t_2 - t_1)$. The remainder of this section will concentrate on compressible flows.

Discharge of Gases

The discharge of gases through orifices or nozzles may be approximated as frictionless adiabatic flow. The reasons for this are as follows: (1) friction losses are minor because of the short distances traveled; and (2) heat transfer is negligible ($\dot{Q} = 0$) because the changes the gas undergoes are slow enough to keep velocity and temperature gradients small. Liepmann and Roshko (1957) provide evidence for these statements.

Applying this approximation, the energy balance expression, Equation (5.67), can be simplified for these flow systems. We shall consider the frequently encountered system of gas discharge from a tank. Due to the relatively low viscosities of gases, flows are often very turbulent and, hence, coefficient α in Equation (5.73) is unity. The height of flow before and after discharge is the same ($Z_1 = Z_2$), and no mechanical work is added or removed from such systems (i.e., $W' = 0$). Thus, Equation (5.73) may be rewritten as follows:

$$i_2 - i_1 = \frac{\overline{w}_2^2}{2g} - \frac{\overline{w}_1^2}{2g} \tag{5.74}$$

where i_1 and i_2 are the enthalpies of 1 kg of gas before and after the expansion, respectively. Thus, $\overline{w}_1$ and $\overline{w}_2$ are the average gas velocities before and after expansion.

As shown by Equation (5.74), the decrease in enthalpy is used to increase the kinetic energy of the system. The linear velocity in a tank, for example before discharge, is insignificant ($w_1 \approx 0$) compared to the velocity at the point of discharge. Therefore, from Equation (5.74),

$$w_2 = \sqrt{2g(i_1 - i_2)} \tag{5.75}$$

And noting that $i_1 - i_2 = C_p(T_1 - T_2)$,

$$w_2 = \sqrt{C_p(T_1 - T_2)2g} \tag{5.76}$$

For reversible adiabatic expansion of an ideal gas, Streeter and Wylie (1979) note the following expression from thermodynamics:

$$\frac{T_1}{T_2} = \left(\frac{P_1}{P_2}\right)^{(\varkappa - 1)/\varkappa} \tag{5.77}$$

where $\varkappa$ is the specific heat ratio: $\varkappa = C_p/C_v$ (C_p denotes heat capacity at constant pressure and C_v at constant volume).

From the ideal gas law ($P_1 v_1' = RT_1$) and the thermodynamic relationship $C_p - C_v = R$, an expression for the velocity of reversible adiabatic discharge of an ideal gas is

$$w_2 = \sqrt{2g \frac{\varkappa}{\varkappa - 1} (P_1 v_1') \left[1 - \left(\frac{P_2}{P_1}\right)^{(\varkappa-1)/\varkappa} \right]} \qquad (5.78)$$

The mass gas rate per unit discharge area may be expressed in terms of the discharge velocity:

$$G = w_2 \gamma_2 = \frac{w_2}{v_2'} \qquad (5.79)$$

where γ_2 is the specific weight of gas, i.e., the inverse of specific volume. Combining Equations (5.79) and (5.78) and including the expression for a reversible adiabatic fluid ($P_1 v_1^\varkappa = P_2 v_2^\varkappa$), the following equation is derived for mass rate in terms of the thermodynamic properties of the fluid:

$$G = \sqrt{2g \frac{\varkappa}{\varkappa - 1} \frac{P_1}{v'} \left[\left(\frac{P_2}{P_1}\right)^{2/\varkappa} - \left(\frac{P_2}{P_1}\right)^{(\varkappa-1)/\varkappa} \right]} \qquad (5.80)$$

Equation (5.80) has a maximum with respect to P_2 or P_2/P_1. Differentiating the expression and setting the derivative equal to zero, the pressure at which the maximum flow rate occurs is obtained. This pressure is referred to as the "critical" pressure:

$$P_{cr} = P_1 \left(\frac{2}{\varkappa + 1}\right)^{\varkappa/(\varkappa-1)} \qquad (5.81)$$

Because $\varkappa$ for gases does not change appreciably, it may be assumed that the critical pressure is in the range of $0.53\, P_1$ to $0.58\, P_1$, i.e., the critical pressure is approximately one-half the tank pressure.

Knowledge of the critical pressure is important for evaluating the efficiency of the flow process. If, for example, the pressure at the exit is higher than the critical value computed from Equation (5.81), then the flow rate in the orifice will not reach its maximum value. At complete expansion, the gas velocity may be computed from Equation (5.78). If the exit pressure of the orifice is less than the critical value, because the maximum flow rate was exceeded, the amount of discharge must reach a maximum value, that is, the critical pressure will be achieved. Thus, further gas expansion will occur downstream of the orifice. The flow will expand, and, consequently, its head will decrease.

Regardless, at $P_2 < P_{cr}$, the gas velocity will not correspond to pressure P_2, and Equation (5.78) should not be used. The discharge velocity will reach a lower value corresponding to P_{cr}. This critical velocity is determined by replacing P_2 in Equation (5.78) with P_{cr}. Taking into account

$$P_1 v_1'^{\varkappa} = P_{cr} v_{cr}'^{\varkappa} \tag{5.82}$$

the critical discharge velocity corresponding to the critical pressure is

$$w_{cr} = \sqrt{g \varkappa P_{cr} v_{cr}'} \tag{5.83}$$

where critical volume v_{cr}' is obtained from Equation (5.82).

Equation (5.83) is an expression for the "sound velocity." Therefore, the maximum linear discharge velocity from a tank orifice is equal to the sound velocity, which is related to a corresponding temperature and pressure at the exit. This temperature may be computed from Equation (5.77) by replacing P_2 and T_2 with P_{cr} and T_{cr}, respectively:

$$T_{cr} = T_1 \left(\frac{2}{\varkappa + 1} \right) \tag{5.84}$$

At any point in the expanding flow the weight velocity is constant and the flow rate changes according to Equation (5.80). Thus, the cross section of the gas flow

$$F = \frac{W}{G} \tag{5.85}$$

will also change. This is illustrated by the plot shown in Figure 5.10. Note that the velocity curve may be computed from Equation (5.78), and the flow rate

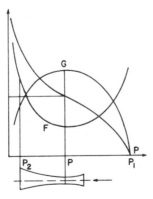

FIGURE 5.10. Plots of flow area and gas mass rate versus pressure for flow through a nozzle.

curve from Equation (5.80), and the section of flow from Equations (5.80) and (5.85).

It is possible to design a nozzle with variable cross section for a specified weight flow rate using Equations (5.80) and (5.85). The critical pressure and velocity will be attained at the narrowest section of the nozzle. In the expanding section, pressure decreases and velocity increases and, according to Equation (5.78), the flow eventually reaches supersonic conditions. The use of a variable cross section nozzle is logical when the counterpressure (downstream) is less than critical pressure P_{cr}. If the downstream pressure exceeds P_{cr}, the effect is still the same because the orifice will have a diameter equal to the largest size of the nozzle's diameter.

Thus far we have considered thermodynamically reversible gas discharge through an orifice or nozzle, i.e., neglecting friction. However, friction can play an important role during discharge in some applications. Friction, associated with the kinetic energy of the gas, is converted into heat, thus increasing the enthalpy of the discharge. Therefore, enthalpy will be less than in the case of reversible flow at the same pressure and counterpressure. Furthermore, Equation (5.75) shows that the discharge gas velocity will decrease. To compute the discharge velocity for this case without considering counterpressure, the temperature of the discharging gas is needed. Knowing the exit temperature T_2 and counterpressure P_2, we find the enthalpy i_2, and the discharge gas velocity is obtained from Equation (5.75). The following example illustrates the use of these equations.

Example 5.6

A reaction takes place in a nitrogen atmosphere in an autoclave at $t = 130°C$ and $p = 10$ atm. The volume of the gas is 100 liters. After 10 minutes it is discovered that the pressure dropped to 9.7 atm because of leakage. The autoclave is found to be rated for pressure only to 6 atm. Determine how long it will take the vessel's pressure to drop to 6 atm and the rate at which nitrogen gas is escaping.

Solution

We will assume that the gas discharge takes place through a narrow slit and is adiabatic and reversible. First, determine the pressure in the autoclave. For nitrogen, $C_v = 5$ cal/mol and $C_p = C_v + R = 7$ cal/mol. Hence,

$$\varkappa = C_p/C_v = 7/5 = 1.4$$

The critical pressure is

$$P_{cr} = P_1\left(\frac{2}{\varkappa + 1}\right)^{\varkappa/(\varkappa-1)} = P_1\left(\frac{2}{1.4 + 1}\right)^{1.4/(1.4-1)} = 0.53P_1$$

We note that the critical pressure will always be greater than the surroundings to which nitrogen is expanding. The gas discharge will occur at the sound velocity, which may be computed from Equation (5.83).

The absolute temperature in the autoclave is $T_1 = 273 + 130 = 430$ K, and, from Equation (5.84), we obtain

$$T_{cr} = 403\left(\frac{2}{1.4 + 1}\right) = 337 \text{ K}$$

The specific volume of nitrogen at T_{cr} and P_{cr} is as follows (molecular weight of nitrogen is 28 and the volume of 1 kg-mol is 22.4 m³):

$$v'_{cr} = \frac{22.4}{28} \times \frac{337}{273} \times \frac{10^4}{0.53P_1} = \frac{18650}{P_1} \text{ m}^3/\text{kg}$$

where 10^4 kg/m² is the atmospheric pressure. Pressure P_1 is expressed in units of kg/m². As noted above, the discharge velocity through the slit may be computed from Equation (5.83):

$$w_{cr} = \sqrt{9.81 \times 1.4 \times 0.53P_1 \times \frac{18650}{P_1}} = 368 \text{ m/s}$$

The nitrogen weight rate at discharge is therefore,

$$G = \frac{w_{cr}}{v'_{cr}} = \frac{368}{18650}P_1 = 0.0197P_1 \frac{\text{kg}}{\text{m}^2\text{-s}}$$

When the pressure is P_1, the specific volume of the nitrogen is as follows:

$$v' = \frac{22.4}{28} \times \frac{403}{273} \times \frac{10^4}{P} = \frac{11850}{P}$$

Because the volume of the autoclave is 100 ℓ (0.1 m³), the weight of nitrogen is

$$W = \frac{0.1P_1}{11850} = 8.42 \times 10^{-6}P$$

The nitrogen loss accompanied by the decrease dp is

$$dW = -8.42 \times 10^{-6}\, dp$$

The leakage occurs when the pressure decreases by some amount dp over a time interval dt through a slit having cross-sectional area F. We can express the discharge by a rate expression:

$$dW = GF dt$$

or

$$-8.42 \times 10^{-6}\, dp = 0.0197\, pF dt$$

Integrating this expression over the pressure limits of 10 atm to P_1, the time over which the pressure drop occurs is obtained:

$$t = -\frac{4.27 \times 10^{-4}}{F} \int_{10}^{P_1} \frac{dP}{dt} = 9.48 \frac{10^{-4}}{F} \log \frac{10}{P_1}$$

where P is expressed in units of absolute atmospheres and t in seconds.

We know from the problem statement that for $t = 10$ min $= 600$ s the pressure in the autoclave is 9.7 atm. From this information, we may compute the area of discharge F in mm²:

$$F = \left(\frac{9.84 \times 10^{-4}}{600} \log \frac{10}{9.7}\right) 10^6 = 0.0213 \text{ mm}^2$$

Hence, the time it takes for the pressure to drop to 6 atm is

$$t = \frac{9.84 \times 10^{-4}}{0.0213 \times 10^{-6}} \log \frac{10}{6} = 10250 \text{ s}$$

Thus, the allowable operating time for the autoclave is about 3 hours. Calculation of the actual weight loss is left to the student.

Gas Flow through Piping

The flow of gases in piping is more complex than that of liquids, mainly because of the dependency of specific weight on pressure changes. Because gases undergo different thermodynamic changes with pressure, specific gravity changes differently. First, we shall consider isothermal flow, i.e., flow at constant temperature. For this case we need to evaluate the amount of heat to be supplied to the flow to maintain a constant gas temperature. We note that the gas pressure decreases because of frictional resistances in the flow direction; hence, the gas specific volume will increase along with the linear velocity and kinetic energy of the gas.

Consider an ideal gas flowing through a section of horizontal piping of constant cross section. The energy balance [Equation (5.73)] may be used to describe any two sections. If the flow is isothermal, then $i_1 = i_2$ and $Z_1 = Z_2$, and the energy balance simplifies to the following:

$$\frac{w_2^2}{2g} - \frac{w_1^2}{2g} = \dot{Q} \qquad (5.86)$$

Equation (5.86) implies that any increase in kinetic energy results from the introduction of heat from the outside. The flow rate can be expressed by

$$w_1 \gamma_1 = w_2 \gamma_2 \qquad (5.87)$$

and the ratio of these velocities for an isothermal system becomes

$$\frac{w_1}{w_2} = \frac{P_2}{P_1} \qquad (5.88)$$

Combining Equations (5.86) and (5.88), we obtain

$$\frac{w_1}{2\gamma}\left[\left(\frac{P_1}{P}\right)^2 - 1\right] = \dot{Q} \qquad (5.89)$$

The above expression may be used to compute the amount of heat introduced to the system, provided we know the velocity at one section and the pressure ratio at the inlet and exit of the flow portion under evaluation.

To determine the pressure drop, a differential form of the Bernoulli equation may be used (the Bernoulli equation is discussed in detail in the following section):

$$v'dp + \frac{dw^2}{2g} + dZ = 0 \qquad (5.90)$$

The change in head dZ may be obtained from the Darcy-Weisbach equation:

$$dZ = \lambda \frac{dL}{D} \times \frac{w^2}{2g} \qquad (5.91)$$

where λ is a flow resistance coefficient.

For a small pressure drop, the specific gas volume may be assumed to be the average over the flow portion:

$$v'_{avg} = \frac{v'_1 + v'_2}{2} = \frac{RT}{2}\left(\frac{1}{P_1} + \frac{1}{P_2}\right) \qquad (5.92A)$$

Similarly, average values for velocity and the resistance coefficient may be defined:

$$w_{avg} = \frac{w_1 + w_2}{2} \qquad (5.92B)$$

$$\lambda_{avg} = \frac{\lambda_1 + \lambda_2}{2} \qquad (5.92C)$$

Combining the Bernoulli and Darcy-Weisbach equations and including the definitions of Equations (5.92A and C), we obtain the following expression for pressure drop:

$$v'_{avg}(P_1 - P_2) = \frac{w_2^2 - w_1^2}{2g} + \lambda_{avg}\frac{L}{D}\frac{w_{avg}^2}{2g} \qquad (5.93)$$

The initial velocity w_1, pressure P_1, and temperature T_1 are usually known. This information may be used to compute the specific volume v'_1 and resistance coefficient λ_1. However, values downstream are often not known but may be solved for by trial and error using the following procedure: Assume a value for P_2 whence specific volume v'_2, velocity w_2 from Equation (5.88), and resistance coefficient λ_2 can be computed. Average values v'_{avg}, w_{avg}, and λ_{avg} may now be calculated. If Equation (5.93) is satisfied, then the initial guess for P_2 was correct. If the equation is not satisfied, then a new value for P_2 must be chosen and the computation repeated until convergence is achieved.

The above discussion was based on the assumption of small pressure drop. If, however, a large pressure drop is expected, the above expressions should be applied over a section of piping of length dL:

$$v'\,dp + \frac{w\,dw}{g} + \lambda\frac{dL}{D}\frac{w^2}{2g} = 0 \qquad (5.94)$$

where

$$\frac{w\,dw}{g} = d\left(\frac{w^2}{2g}\right)$$

Dividing Equation (5.94) by v'^2 and noting that $G = W/v'$ and $Pv' = RT$, we obtain the following expression after integration:

$$-\frac{1}{R}\int_{P_1}^{P_2}\frac{pdp}{T} = \frac{G^2}{g}\int_{v_1'}^{v_2'}\frac{dv'}{v'} + \frac{G^2}{2gD}\int_0^L \lambda dL \qquad (5.95)$$

The resistance coefficient λ for isothermal gas flow through a constant cross section must be constant, because it is a function of the Reynolds number ($Re = DG/\mu g$), which is also constant for a specified flow rate.

On integration of Equation (5.95) we obtain

$$\frac{P_1^2 - P_2^2}{2RT} = \frac{G^2}{g}\ln\frac{v_2'}{v_1'} + \frac{\lambda L G^2}{2gD} \qquad (5.96)$$

Equation (5.96) permits us to determine the pressure drop $P_1 - P_2$ along a piping section of length L by a method of successive approximations. Let us assume that specific volume v_2' depends on unknown pressure P_2. Denoting $\gamma = (P_1 + P_2)/2RT$ as a mean arithmetic value of v_1' and v_2', Equation (5.96) may be simplified to the following form:

$$P_1 - P_2 = \frac{G^2}{g\gamma}\ln\frac{v_2'}{v_1'} + \frac{\lambda G^2 L}{2gD\gamma} \qquad (5.97)$$

This last expression is also applicable to nonisothermal flows. Now it is convenient to use the average specific volume from Equation (5.92A) rather than the average specific weight in the Bernoulli equation:

$$P_1 - P_2 = \frac{G^2}{g}(v_1' - v_2') = \frac{\lambda G^2 L}{2gD\gamma} \qquad (5.98)$$

Solution of Equations (5.98), (5.97), and (5.96) can be obtained by the method of successive approximations.

We will now develop an expression for the maximum isothermal flow rate. Rearranging Equation (5.97) to solve for G,

$$G = \sqrt{\frac{(P_1^2 - P_2^2)g}{2RT\left(\ln\frac{P_1}{P_2} - \frac{\lambda L}{D}\right)}} \qquad (5.99)$$

If P_2 changes, the flow rate maximizes at some critical value of pressure P_{cr}.

The new critical pressure may be determined by differentiating Equation (5.99) with respect to P_2 and setting the derivative equal to zero:

$$\ln \frac{P_1}{P_{cr}} + \frac{\lambda L}{D} = \frac{P_1^2 - P_{cr}^2}{2P_{cr}^2} \tag{5.100}$$

The critical pressure from this expression is substituted into Equation (5.99) to obtain the maximum flow rate:

$$G_{max} = \sqrt{\frac{P_{cr}^2 g}{RT}} \tag{5.101}$$

Noting that $v'_{cr} = RT/P_{cr}$, Equation (5.101) can be rearranged to give the maximum velocity for pipe flow:

$$w_{max} = \sqrt{gP_{cr}v'_{cr}} \tag{5.102}$$

Thus, there is a limiting gas velocity in piping that corresponds to the critical pressure at the exit. If pressure falls below this value, the gas velocity will not increase.

Let us now consider another limiting case—that of frictionless adiabatic flow. That is, we will assume the pipe system to be perfectly thermally insulated. The energy balance equation simplifies to the following form:

$$i_1 - i_2 = \frac{w_2^2 - w_1^2}{2g} \tag{5.103}$$

or

$$C_p(T_2 - T_1) = \frac{w_1^2}{2g}\left[\left(\frac{T_2 P_1}{T_1 P_2}\right)^2 - 1\right] \tag{5.104}$$

This last expression is developed from Equation (5.103) using $i_1 - i_2 = C_p(T_2 - T_1)$ for perfect gases and $w_2/w_1 = T_2 P_1/T_1 P_2$ from $pv' = RT$ and $w_2/w_1 = v'_2/v'_1$. In practice, because of frictional resistances $P_1 > P_2$, and, consequently, a change in temperature does occur.

To develop an expression for pressure drop, the method of Lapple (1947) is used. The derivation is based on the Bernoulli and Darcy-Weisbach equations expressed in differential form:

$$v'dp + \frac{wdw}{g} + \lambda \frac{dL}{D} \times \frac{w^2}{2g} = 0 \tag{5.105}$$

denoting

$$v' dp = d(pv') - pv' \times \frac{dv'}{v'} \qquad (5.106)$$

And, for the energy balance equation,

$$C_p dT = -d\left(\frac{w^2}{2g}\right) \qquad (5.107)$$

where $dT = d(pv')/R$ and $R = C_p - C_v$.

Equation (5.107) may be rewritten as

$$d(pv') = -\left(\frac{\varkappa - 1}{\varkappa}\right) d\left(\frac{w^2}{2g}\right) \qquad (5.108)$$

Substituting Equation (5.108) into Equation (5.106) and then into Equation (5.105), we obtain the following:

$$(1 + \varkappa)\frac{dw}{w} - [2g\varkappa P_1 v_1' + (\varkappa - 1)w_1^2]\frac{dw}{w^3} + \frac{\varkappa \lambda}{D} dL = 0 \qquad (5.109)$$

According to Equation (5.83), the velocity of sound C_1 at T_1 corresponding to P_1 and v_1' is

$$C_1 = \sqrt{g \varkappa P_1 v_1'} \qquad (5.110)$$

Hence, Equation (5.109) may be rewritten as

$$(1 + \varkappa)\frac{dw}{w} - [2C_1^2 + (\varkappa - 1)w_1^2]\frac{dw}{w^3} + \frac{\varkappa \lambda}{D} dL = 0 \qquad (5.111)$$

On integration, we obtain

$$\lambda \frac{L}{D} = -\left(\frac{\varkappa + 1}{\varkappa}\right)\ln\frac{w_1}{w_2} + \frac{1}{\varkappa}\left(\frac{C_1^2}{w_1^2} + \frac{\varkappa + 1}{2}\right)\left(1 - \frac{w_1^2}{w_2^2}\right) \qquad (5.112)$$

If the initial gas velocity w_1 is known, w_2 may be computed from this expression for a specified pipe length L. The value of λ does not change significantly and should be taken only as an average value when considering long lengths of piping. From w and w_1, Equations (5.103) and (5.109) are combined to give

$$\frac{T_2}{T_1} = 1 + \frac{\varkappa - 1}{2C_1^2} w_1^2 \left(1 - \frac{w_2^2}{w_1^2}\right) \qquad (5.113)$$

and, from the equation of continuity,

$$\frac{P_2}{P_1} = \frac{w_1}{w_2}\left[1 + \frac{\varkappa - 1}{2C_1^2}w_1^2\left(1 - \frac{w_2^2}{w_1^2}\right)\right] \qquad (5.114)$$

Analysis of the above equations for adiabatic gas flow in piping reveals that there is a maximum flow rate where the gas velocity at the exit reaches the velocity of sound. However, the adiabatic gas flow expressions provide essentially the same results as does the isothermal analysis. For very short pipes and high-pressure gradients, the adiabatic flow rate will be larger than the isothermal case; however, differences generally are no greater than 20%. Further discussions are given by Cambel and Jennings (1958), Shapiro (1958), Liepmann and Roshko (1957). The following three examples illustrate applications of the flow analyses presented. Additional problems for the student are given at the end of this chapter.

Example 5.7

Air is flowing through a horizontal pipeline at a rate of 280 kg/hr. The pipe is 52.5 mm (2 in) in diameter and 150 m in length. The exit pressure is atmospheric. Compute the pressure drop in the piping. The flow may be assumed isothermal at $T = 20°C$.

Solution

The cross section of the piping is

$$F = \frac{\pi \times 0.0525^2}{4} = 2.17 \times 10^{-3} \text{ m}^2$$

The mass flow rate is

$$W = \frac{280}{3600} = 7.76 \times 10^{-2} \text{ kg/s}$$

Hence, the specific weight flow rate is

$$G = \frac{W}{F} = \frac{7.76 \times 10^{-2}}{2.17 \times 10^{-3}} = 35.9 \frac{\text{kg}}{\text{m}^2 \times \text{s}}$$

The viscosity of air at 20°C is 0.02 cp, i.e., $\mu_g = 0.02 \times 10^{-3}$ kg/m-s. We now have enough information to compute the Reynolds number:

$$Re = \frac{GD}{\mu_g} = \frac{35.9 \times 0.0525}{0.02 \times 10^{-3}} = 94,200$$

The friction coefficient may be computed from the following turbulent correlation given by Perry (1950):

$$\lambda = 0.0123 + \frac{0.7544}{Re^{0.38}} = 0.0123 + \frac{0.754}{94,200^{0.38}} = 0.022$$

The specific volume of the gas is

$$v_2' = \frac{22.4}{29} \times \frac{(273 + 20)}{273} = 0.83 \text{ m}^3/\text{kg}$$

We can now use Equation (5.98) to compute the upstream pressure:

$$P_1 = 10,333 + \frac{35.9^2}{9.81\gamma} \ln \frac{0.83}{1} + \frac{0.022 \times 359^2}{2 \times 9.81 \times 0.0525\gamma}$$

or

$$P_1 = 10,333 + \frac{302}{\gamma} \log \frac{0.83}{v_1'} + \frac{4120}{\gamma}$$

This expression may be solved by a trial and error solution, i.e., a method of successive approximations.

The gas constant $R = 1.987$ kcal/mol K, and 1 kcal $= 426.7$ kg-m. Therefore,

$$R = \frac{1.987 \times 426.7}{2g} = 29.2 \frac{\text{kg-m}}{\text{kg-K}}$$

$$T = 273 + 20 = 293 \text{K}$$

$$RT = 29.2 \times 293 = 8550 \text{ m}$$

$$v_1' = \frac{RT}{P_1} = \frac{8550}{P_1}$$

The average specific volume is

$$\gamma = \frac{P_1 + P_2}{2RT} = \frac{P_1 + 10{,}333}{17{,}100}$$

Assuming a value for P_1 of 13,000 kg/m².

$$v_1' = \frac{8{,}550}{13{,}000} = 0.66 \text{ m}^3/\text{kg}$$

and the average specific weight is

$$\gamma = \frac{13{,}000 + 10{,}333}{17{,}100} = 1.36 \text{ kg/m}^3$$

Substituting these values into our equation,

$$P_1 = 10{,}333 + \frac{302}{1.36} \log \frac{0.83}{0.66} + \frac{4120}{1.36} = 13{,}380 \text{ kg/m}^2$$

This computed value is a little different from the assumed P_1 (13,000) and, hence, a new P_1 should be selected and the calculations repeated. To expedite computations, the equations were programmed in BASIC format with a desktop computer. The program is listed in Table 5.1, and tabulated values are given in Table 5.2. The program translated computations to a printer, which produced a plot of assumed P_1 versus computed P_1, as shown in Figure 5.11. The intersection of the curve generated by calculations with the bisectrix, i.e., the line of perfect agreement, provides the solution ($P_1 = 13{,}350$ kg/m²).

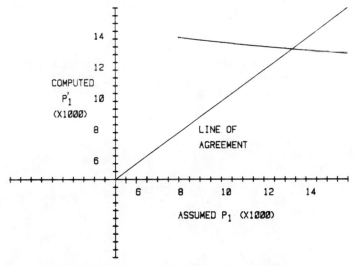

FIGURE 5.11. Plot of computed versus assumed values of P_1.

Table 5.1. Computer Program to Solve for Upstream Pressure in Example 5.7.

```
10 DIM P1(15),N1(15),G(15),C(15
   )
20 READ T,R,P2
30 P1(1)=8000
40 REM COMPUTES P1 VALUES
50 PRINT " ASMD.    CMPTD."
60 PRINT "   P1       P1      NU
   GAMMA"
70 PRINT "_____
   _____"
80 FOR I=1 TO 10
90 N1(I)=R*(T+273)/P1(I)
100 G(I)=(P1(I)+P2)/(2*R*(T+273)
    )
110 H=302/G(I)*LOG(.83/N1(I))
120 C(I)=P2+H+4120/G(I)
130 PRINT USING 140 ; P1(I),C(I)
    ,N1(I),G(I)
140 IMAGE DDDDDD,DDDDDDDD,DDD.DD
    D,DDDD.DD
150 PRINT
160 P1(I+1)=P1(I)+1000
170 P1(I)=P1(I)/1000
180 C(I)=C(I)/1000
190 NEXT I
200 DATA 20,29.2,10333
210 REM DRAWS PARITY PLOT
220 GCLEAR
230 SCALE 0,16,0,16
240 XAXIS 5,.5
250 YAXIS 5,.5
260 MOVE 5,5
270 DRAW 16,16
280 FOR I=1 TO 7
290 MOVE P1(1),C(1)
300 FOR I=2 TO 10
310 DRAW P1(I),C(I)
320 NEXT I
330 LDIR 0
340 FOR X=6 TO 16 STEP 2
350 MOVE X,4
360 LABEL VAL$(X)
370 NEXT X
380 LDIR 0
390 FOR Y=6 TO 16 STEP 2
400 MOVE 4,Y
410 LABEL VAL$(Y)
420 NEXT Y
430 LDIR 0
440 MOVE 6,2.5
450 LABEL "ASSUMED P1
    (X1000)"
460 LDIR 0
470 MOVE  5,11
480 LABEL "COMPUTED"
490 MOVE 1.9,10
500 LABEL "P1"
510 MOVE 5,9
520 LABEL "(X1000)"
530 LDIR 0
540 MOVE 9,8
550 LABEL "LINE OF"
560 MOVE 9,7
570 LABEL "AGREEMENT"
580 COPY
590 END
```

Table 5.2. Tabulated Values Computed for Example 5.7.

Assumed P_1	Computed P_1	v'	γ
8,000	14,107	1.069	1.07
9,000	13,943	0.951	1.13
10,000	13,792	0.856	1.19
11,000	13,653	0.778	1.25
12,000	13,525	0.713	1.31
13,000	13,406	0.658	1.36
14,000	13,295	0.611	1.42
15,000	13,192	0.570	1.48
16,000	13,096	0.535	1.54
17,000	13,007	0.503	1.60

Hence, the pressure drop is

$$P_1 - P_2 = 13{,}350 - 10{,}333$$

$$= 3017 \text{ kg/m}^2 \text{ (or 0.3 atm)}$$

The student should prepare a separate program with a convergence routine to obtain the pressure drop. Specify a tolerance, i.e., the agreement between assumed and computed P_1, of 1%.

Example 5.8

Determine the upstream pressure for methane flowing at a rate of 1.2 kg/s in a pipe with a 130-mm i.d. and a 30-km length. Conditions at the exit are 2.5 atm and 20°C.

Solution

For isothermal flow of a real gas, the increase in internal energy equals zero and density is constant. Hence, the equation for a differential element of piping is as follows:

$$\frac{d(w^2)}{2} + \frac{dP}{\varrho} + \delta\hat{F} = 0 \qquad \text{(i)}$$

where $\hat{F}$ is the friction energy per unit mass.

The change in sign on $\hat{F}$ is associated with the inverse of terms containing infinitesimal differences. Velocity may be expressed as follows:

$$w = \frac{4G}{\pi d^2 \varrho}$$

or

$$d(w^2) = \left(\frac{4G}{\pi d^2}\right) 2v' dv'$$

where v = specific volume. Substituting this relationship into Equation (i), we obtain

$$\left(\frac{4G}{\pi d^2}\right)^2 \frac{dv'}{v'} + \frac{dp}{v'} + \frac{\lambda}{2}\left(\frac{4G}{\pi d^2}\right)^2 \frac{d\ell}{d} = 0 \qquad \text{(ii)}$$

To integrate Equation (ii), the relationship of pressure, specific volume, and λ must be known.

For perfect gases at isothermal conditions,

$$Pv' = \text{constant} = P_2 v_2'$$

and

$$\lambda = \text{constant because } Re = \frac{wd}{\nu} = \frac{G}{\pi d \mu} = \text{constant}$$

Substituting this expression in Equation (ii), we obtain the following:

$$\left(\frac{4G}{\pi d^2}\right)^2 \frac{dP}{P} + \frac{PdP}{P_2 v_2'} + \frac{\lambda}{2}\left(\frac{4G}{\pi d^2}\right)^2 \frac{d\ell}{d} = 0$$

Integrating over the limits of 0 to ℓ, between P_1 and P_2, we obtain

$$\left(\frac{4G}{\pi d^2}\right)^2 \left(\ln \frac{P_1}{P_2} + \frac{\lambda}{2}\frac{\ell}{d}\right) = \frac{P_1^2 - P_2^2}{P_2 v_2'} \tag{iii}$$

To evaluate the friction factor, we first compute the Reynolds number:

$$Re = \frac{4G}{\pi d \mu} = \frac{4 \times 1.2}{\pi \times 0.13 \times 1.08 \times 10^{-5}} = 1.09 \times 10^6$$

Using a roughness coefficient $e = 0.1$ mm, and relative roughness $e/d = 0.1/130 = 0.00077$, we obtain from the Moody plot a value for λ:

$$\lambda = 0.018$$

The specific volume of methane at 2.5 atm, 20°C, is

$$v_2' = \frac{22.4}{16} \times \frac{1.033}{2.5} \times \frac{273 + 20}{273} = 0.62 \text{ m}^3/\text{kg}$$

Substituting these values into Equation (iii), we have

$$\left(\frac{4 \times 1.2}{\pi \times 0.13^2}\right)^2 \left(\ln \frac{P_1}{2.5 \times 9.81 \times 10^4} + \frac{0.018 \times 30{,}000}{2 \times 0.13}\right)$$

$$= \frac{P_1^2 - (2.5 \times 9.81 \times 10^4)^2}{2.5 \times 9.81 \times 10^4 \times 0.62}$$

To simplify calculations, we will neglect the term $\ln P_1/P_2$, assuming it to be small in comparison to $\lambda/2 \times l/d$. Hence,

$$P_1 = \sqrt{(2.5 \times 9.81 \times 10^4)^2 + 2.5 \times 9.81 \times 10^4 \times 0.62\left(\frac{4 \times 1.2}{\pi \times 0.13^2}\right)^2 \frac{0.018 \times 30{,}000}{2 \times 0.13}}$$

$$= 1.625 \times 10^6 \text{ N/m}^2 = 16.6 \text{ atm}$$

By checking the error obtained as a result of this assumption, we find:

$$\ln \frac{P_1}{P_2} = 1.89 \ll \frac{\lambda}{2} \cdot \frac{\ell}{d} = \frac{0.018 \times 30{,}000}{2 \times 0.13} = 2075$$

Hence, the calculation based on the initial pressure is sufficiently accurate.

Example 5.9

Determine the optimum pipe diameter for transporting 6000 N·m³/hr of methane over a distance of 4 km. The efficiency of the gas blower is 0.5 and electrical costs are 50 ¢/kWh. The amortization cost of piping is $24/m·yr for 1-m i.d. pipe. Maintenance costs are $18/m·yr.

Solution

Assume that $\lambda = 0.03$ and local losses are 10% of the friction losses. The calculation will be based on a temperature of 30°C.

$$V_{sec} = \frac{6000 \times 303}{3600 \times 273} = 1.85 \text{ m}^3/\text{s}$$

$$w = \frac{V_{sec}}{0.785 \, d^2} = \frac{1.85}{0.785 \, d^2} = \frac{2.36}{d^2} \text{ m/s}$$

$$\Delta p = \Delta p_f + \Delta p_{\ell.\ell.} = 1.1 \, \Delta p_f$$

Hence, the pressure losses are

$$\Delta p = \frac{1.1 \, \lambda L w^2 \gamma}{d \times 2g} = \frac{1.1 \times 0.03 \times 4000 \times 2.36^2 \times 0.64}{d \times 2 \times 9.81 \times d^4}$$

$$= \frac{24}{d^5} \text{ kg/m}^2$$

The specific weight of methane is

$$\gamma = \frac{16 \times 273}{22.4 \times 303} = 0.64 \text{ kg/m}^3$$

We calculate the consumed power assuming that $\Delta p < 0.1$ atm (we will check it at the end of the design):

$$N = \frac{V_{sec}\Delta p}{102\eta} = \frac{1.85 \times 24}{102 \times 0.5 \times d^5} = \frac{0.87}{d^5} \text{ kW}$$

One kW-yr costs $0.4 \times 24 \times 330 = \$3160$, assuming 330 working days. Thus, the cost of energy is

$$E = \frac{0.97 \times 3160}{d^5} = \frac{2750}{d^5} \text{ \$/yr}$$

The amortization costs are

$$A = 24 \times L \times d = 24 \times 4000 \times d = 96{,}000\, d \text{ \$/yr}$$

Maintenance costs are

$$M = 18 \times L \times d = 72{,}000\, d \text{ \$/yr}$$

The total cost as a function of pipe diameter is

$$E + A + M = \frac{2750}{d^5} + 168{,}000 \text{ \$/yr}$$

To determine the minimum costs (and, hence, the optimum diameter, assuming the Δp is acceptable), we differentiate this expression and equate it to zero:

$$\frac{\partial}{\partial d}(E + A + M) = -5 \times 2750\, d^{-6} + 168{,}000 = 0$$

Hence,

$$d = 0.66 \text{ m}$$

Finally, we check the pressure drop:

$$\Delta p = \frac{24}{d^5} = \frac{24}{0.66^5} = 193 \text{ kg/m}^2 = 0.0193 \text{ atm}$$

The pressure drop is less than the 0.1 atm assumed.

THE BERNOULLI EQUATION

General Application

In this subsection, a special case of the total energy balance is considered. Integration of Euler's equations of motion for steady flow leads to one of the most important and widely used expressions, namely, the Bernoulli equation. Multiplying both sides of the system of Equations (5.51) by dx, dy, and dz, respectively, and then dividing by density ϱ, we obtain the following:

$$\frac{dx}{dt} dw_x = -\frac{1}{\varrho} \frac{\partial P}{\partial x} dx$$

$$\frac{dy}{dt} dw_y = -\frac{1}{\varrho} \frac{\partial P}{\partial y} dy$$

$$\frac{dz}{dt} dw_z = -gdz - \frac{1}{\varrho} \frac{\partial P}{\partial z} dz$$

Summing these equations and noting that the derivatives dx/dt, dy/dt, and dz/dt are velocity projections w_x, w_y, and w_z on the corresponding coordinate axes, we obtain

$$w_x dw_x + w_y dw_y + w_z dw_z + - gdz - \frac{1}{\varrho}\left(\frac{\partial P}{\partial x} dx + \frac{\partial P}{\partial y} dy + \frac{\partial P}{\partial z} dz\right)$$

(5.115)

The terms on the LHS are

$$w_x dw_x = d\left(\frac{w_x^2}{2}\right); \quad w_y dw_y = d\left(\frac{w_y^2}{2}\right); \quad w_z dw_z = d\left(\frac{w_z^2}{2}\right)$$

Hence, their sum may be written as

$$d\left(\frac{w_x^2}{2}\right) + d\left(\frac{w_y^2}{2}\right) + d\left(\frac{w_z^2}{2}\right) = d\left(\frac{w_x^2 + w_y^2 + w_z^2}{2}\right) = d\left(\frac{w^2}{2}\right) \quad (5.116)$$

where $w = |\overline{W}|$ is the velocity vector with components along coordinate axes w_x, w_y, and w_z.

The sum of the terms in parentheses on the RHS of Equation (5.115) is the total pressure differential dp. Recall that the pressure at any one point does not change with time, i.e., only spacial coordinates are considered. Hence, we may write

$$d\left(\frac{w^2}{2}\right) = -\frac{dp}{\varrho} - dz \qquad (5.117)$$

Dividing both sides by g and specifying incompressible homogeneous flow, i.e., ϱ is constant, we obtain

$$d\left(\frac{w^2}{2g}\right) + \frac{dp}{\varrho g} + dz = 0 \qquad (5.118)$$

The sum of the differential then is substituted by the differential of the sum

$$d\left(z + \frac{P}{\varrho g} + \frac{w^2}{2g}\right) = 0$$

Hence,

$$z + \frac{P}{\varrho g} + \frac{w^2}{2g} = \text{constant} \qquad (5.119)$$

This expression may be related to any two cross sections of a flow element in the following manner:

$$z_1 + \frac{P_1}{\varrho g} + \frac{w_1^2}{2\alpha g} = z_2 + \frac{P_2}{\varrho g} + \frac{w_2^2}{2\alpha g} \qquad (5.120)$$

Note that velocities and pressures are not equivalent at different points at any cross section (refer back to the velocity profiles shown in Figure 5.4). Equation (5.120) is related not to the cross section as a whole, but to any pair of compatible points in these sections (for example, points located along the axis of the piping). To compare congruent values over total cross sections, it must be assumed that the terms in Equation (5.38) may be approximated by Equation (5.119), which represents the point form of the *Bernoulli equation for ideal fluids*.

It follows that the *hydrodynamic head* ($z + P/\varrho g + w^2/2g$) is constant for all cross sections of ideal steady flow. The first two terms of the hydrodynamic head, z and $P/\varrho g$, were introduced in Equation (4.7). The *leveling height*, also called the *geometric head*, is z, which represents a specific potential energy at a given point within the cross section. $P/\varrho g$ is the static or piezometric head characterizing the specific potential energy of pressure at a given point. Both terms may be expressed in units either of length or specific energy, i.e., energy per unit weight of fluid.

The third item $w^2/2g$ is also expressed in units of length:

$$\left[\frac{w^2}{2g}\right] = \left[\frac{m^2 \times s^2}{s^2 \times m}\right] = [m]$$

or, after multiplying and dividing by unit weight, N, in units of energy:

$$\left[\frac{w^2}{2g}\right] = \left[\frac{N \times m}{N}\right] = \left[\frac{J}{N}\right]$$

$w^2/2g$ is called the *velocity or dynamic head*, which characterizes a specific kinetic energy at a given point within the cross section.

Thus, according to the Bernoulli equation, the sum of the static and dynamic heads and level do not change from one cross section of flow to another. This statement is inclusive only of steady-state flow of an ideal fluid. Furthermore, it follows that the sum of the potential and kinetic energies is constant which is further illustrated by the following discussions.

A conversion of energy occurs when the flow changes. This change in energy is reflected in the fluid velocity. Reducing a pipe's cross-sectional area causes part of the potential energy of pressure to be converted into kinetic energy and vice versa. When the flow area is increased, part of the kinetic energy is converted into potential energy; however, the total energy is unchanged. Hence, we may conclude that the amount of energy entering the initial pipe cross section is equal to that leaving the pipe. For this reason, Equation (5.120) is a special application of the principle of conservation of energy.

Multiplying Equation (5.120) by the specific weight $\gamma = \varrho g$, we obtain

$$\varrho g z_1 + P_1 + \frac{\varrho w_1^2}{2} = \varrho g z_2 + P_2 + \frac{\varrho w_2^2}{2} \qquad (5.121)$$

where each term expresses the specific energy of the flow per unit volume at a given point. For example,

$$[P] = \left[\frac{N}{m^2}\right] = \left[\frac{N \times m}{m^2 \times m}\right] = \left[\frac{J}{m^2}\right]$$

For a horizontal system, $z_1 = z_2$ and Equation (5.121) simplifies to

$$\frac{P_1}{\varrho g} + \frac{w_1^2}{2g} = \frac{P_2}{\varrho g} + \frac{w_2^2}{2g} \qquad (5.122)$$

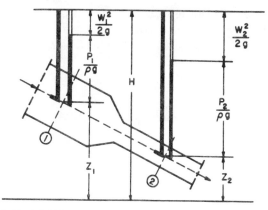

FIGURE 5.12. Ideal fluid flowing through a variable cross section located arbitrarily in space.

To illustrate an application of the Bernoulli equation, consider the variable cross-sectional flow system illustrated in Figure 5.12. Points on the flow axis at sections 1 and 2 are at heights z_1 and z_2 above the datum, respectively. At each of these points, two piezometric tubes are inserted into the flow. One tube at each point has its end into the direction of flow. The levels of fluid in the straight vertical tubes settle at heights corresponding to the hydrostatic pressures at the points of their submergence, i.e., they provide a measure of the *static head* at corresponding points. The fluid height in the bent tubes is higher than in the straight ones because the measurement represents the *sum of the static and dynamic heads*. According to Equation (5.119), the levels in the bent tubes are the same, because they are both referenced to the same datum plane and are a measure of the hydrodynamic head.

As the flow cross section at plane 2-2 is less than that of 1-1 from continuity, fluid velocity w_2 (for a constant flow rate) must be greater than w_1. Hence, the kinetic energy at 2-2 is greater than at 1-1 ($w_2^2/2g > w_1^2/2g$). Therefore, the difference between static and dynamic heads at plane 2-2 is greater than the difference at plane 1-1.

The Bernoulli equation states that the fluid level in the straight tube at 2-2 is less than the corresponding level in the straight tube at plane 1-1 and, by the same value, that the velocity head at 2-2 exceeds that at 1-1. This example illustrates the mutual conversion of potential energy into kinetic when the flow cross section is changed. The overall conclusion is that the sum of these energies in any cross section of the piping remains unchanged.

For real fluids, both shear and friction forces play important roles. These latter forces exert resistance to fluid motion. A portion of the flow energy must be devoted to overcome this *hydraulic resistance*. The total energy decreases continuously in the direction of flow as a portion is converted from potential energy into *lost energy* (the energy expended for friction). This conversion is irreversible and is lost in the form of heat dissipation to the surroundings. For the system just analyzed, this means that

$$z_1 + \frac{P_1}{\varrho g} + \frac{w_1^2}{2g} > z_2 + \frac{P_2}{\varrho g} + \frac{w_2^2}{2g}$$

Thus, for real (viscous) fluids, the levels in the bent tubes at planes 1-1 and 2-2 in Figure 5.12 are not the same. The difference in levels is attributed to energy losses in the fluid path from 1-1 to 2-2 and is referred to as the *lost head* h_ℓ. The Bernoulli equation may be corrected for this frictional loss by adding the term h_ℓ to the RHS of Equation (5.120). Hence, the Bernoulli equation for real fluids is written as follows:

$$z_1 + \frac{P_1}{\varrho g} + \frac{w_1^2}{2g} = z_2 + \frac{P_2}{\varrho g} + \frac{w_2^2}{2g} + h_\ell \qquad (5.123)$$

where the lost head h_ℓ, characterizes the *specific energy spent for overcoming hydraulic resistances*.

Another form of this expression can be obtained by multiplying both sides by ϱg:

$$\varrho g z_1 + \frac{\varrho w_1^2}{2\alpha} + P_1 = \varrho g z_2 + P_2 + \frac{\varrho w_2^2}{2\alpha} + \Delta P_\ell \qquad (5.124)$$

ΔP_ℓ is the lost pressure drop defined as

$$\Delta P_\ell = \varrho g h_\ell \qquad (5.125)$$

The following examples apply to the above principles.

Example 5.9

Oil (specific gravity 0.89) is being pumped through a pipe system of constant diameter. The pressure just upstream of the pump is 25 kN/m² abs. and the pump's discharge pressure is 73 kN/m² abs. The discharge pipe of the pump is 8 m above the centerline of the inlet pipe. The pump supplies 180 J/kg of fluid flowing in the pipe. The flow through the pipe system was determined to be turbulent. Compute the frictional losses in the system.

Solution

We shall choose the centerline of the pipe inlet as the reference datum plane. Hence, $z_1 = 0$ and $z_2 = 8$ m. As the pipe diameters upstream and downstream of the pump are the same, $w_1 = w_2$.

This is a steady-state, incompressible flow problem involving application of the total energy equation, Equation (5.123). As the flow is turbulent, no correction is needed for the kinetic energy terms, i.e., $\alpha = 1.0$. Rewritting Equation (5.123) to the following to solve for friction losses:

$$\Sigma h_\ell = -\hat{W}_s + \frac{1}{2\alpha}(w_1^2 - w_2^2) + g(z_1 - z_2) + \frac{P_1 - P_2}{\varrho}$$

where $\hat{W}_s$ is the work performed by the pump, and Σh_ℓ represents the sum of the friction losses, i.e., head losses. Evaluating each term,

$$\frac{1}{2(1)}(w_1^2 - w_2^2) = 0$$

$$g(z_1 - z_2) = (9.8 \text{ m/s}^2)(0 - 8) \text{ [m]} = -78.4 \text{ J/kg}$$

$$\frac{P_1 - P_2}{\varrho} = \frac{25 - 73 \text{ [kN/m}^2\text{]}}{890 \text{ kg/m}^3} \times 1000 = -53.9 \text{ J/kg}$$

Hence,

$$\Sigma h_\ell = -(-180) + 0 + (-78.4) + (-53.9)$$

$$= 47.7 \text{ J/kg}$$

or

$$\Sigma h_\ell = 47.7 \times 0.33485 = 15.96 \frac{\text{ft-lb}_f}{\text{lb}_m}$$

Example 5.10

A liquid is being pumped from an open tank to a height of 37 ft above the initial level in the tank. The density of the liquid is 73 lb_m/ft^3 and the pumping rate is 35 gpm. The discharge line on the pump is 2.5 in i.d. The total friction loss in the piping is 23 $\text{ft-lb}_f/\text{lb}_m$. The level in the tank is dropping at a rate of 0.25 fps, and the pump's rated efficiency is 59%. Assuming flow through the piping is turbulent, determine the horsepower of the pump.

176 HYDRODYNAMICS: SINGLE-FLUID FLOWS

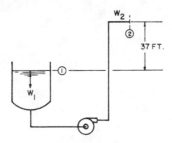

FIGURE 5.13. Flow system under evaluation in Example 5.10.

Solution

The flow diagram for this system is given in Figure 5.13. The mechanical energy actually delivered to the fluid by the pump, i.e., the net mechanical work, is

$$\hat{W}_s = -\eta W_p$$

where η is the fractional efficiency of the pump, and W_p is the shaft work delivered to the pump.

The volumetric flow rate is

$$V = \left(35 \, \frac{\text{gal}}{\text{min}}\right)\left(\frac{1 \, \text{min}}{60 \, \text{s}}\right)\left(\frac{\text{ft}^3}{7.481 \, \text{gal}}\right) = 0.07798 \text{ cfs}$$

The fluid velocity in the tank is given, $W_1 = 0.25$ fps. The cross section of the discharge pipe is

$$F = \frac{1}{4} \pi (2.5/12)^2 = 0.0341 \text{ ft}^2$$

Hence, the velocity downstream of the pump is

$$w_2 = \frac{0.07798 \text{ cfs}}{0.0341 \text{ ft}^2} = 2.288 \text{ fps}$$

We will assume that the discharge is open to the atmosphere. Hence,

$$P_1 = P_2 = 1 \text{ atm}$$

Therefore,

$$\frac{P_1}{\varrho} - \frac{P_2}{\varrho} = 0$$

And because $\alpha = 1.0$,

$$\frac{w_1^2}{2g_c} = \frac{(0.25 \text{ fps})^2}{2(32.174)} = 9.713 \times 10^{-4} \frac{\text{ft-lb}_f}{\text{lb}_m}$$

$$\frac{w_2^2}{2g_c} = \frac{(2.288 \text{ fps})^2}{2(32.174)} = 8.135 \times 10^{-2} \frac{\text{ft-lb}_f}{\text{lb}_m}$$

Assigning the initial level in the tank as the reference datum plane, we note that

$$z_1 = 0$$

and

$$z_2 \frac{g}{g_c} = (37.0 \text{ ft})\left(\frac{32.2}{32.174}\right) = 37.0 \frac{\text{ft-lb}_f}{\text{lb}_m}$$

We now rearrange the mechanical energy equation to solve for mechanical work:

$$\hat{W}_s = \frac{g}{g_c}(z_1 - z_2) + \frac{1}{2g_c}(w_1^2 - w_2^2) + \frac{P_1 - P_2}{\varrho} - \Sigma \hat{F}$$

$$\hat{W} = -37.0 + (9.713 \times 10^{-4} - 8.135 \times 10^{-2}) + 0 - 23$$

$$= -60.08 \frac{\text{ft-lb}_f}{\text{lb}_m}$$

Hence,

$$\hat{W}_p = -\frac{\hat{W}_s}{\eta} = -\frac{(-60.08)}{0.59} = 101.83 \frac{\text{ft-lb}_f}{\text{lb}_m}$$

Mass flow rate = $(0.07798 \text{ cfs})(73 \text{ lb}_m/\text{ft}^3) = 5.69 \text{ lb}_m/\text{s}$. The pump horsepower therefore is

$$\left(5.69 \frac{\text{lb}_m}{\text{s}}\right)\left(101.83 \frac{\text{ft-lb}_f}{\text{lb}_m}\right)\left(\frac{1 \text{ hp}}{550 \frac{\text{ft-lb}_f}{\text{s}}}\right) = 1.05 \text{ hp}$$

Additional examples are given by Bird et al. (1960), Cheremisinoff (1981a,b), Mott (1972), and Pai (1956).

Variable-Head Meters

The determination of head losses is an important practical problem connected with the calculation of the energy required for fluid displacement in pumps, compressors, etc., as well as for measuring flow quantities entering and leaving process equipment. Many flow problems in this category can be addressed through the Bernoulli equation. We will discuss some practical applications of the Bernoulli equation through the use of variable-head meters. Such devices are used extensively throughout industry to measure and control flows through equipment.

However, we shall limit our discussion to the three most widely used head meters: pitot tubes, orifice and venturi meters. Extensive data on the design of these devices are given by the American Society of Mechanical Engineers [ASME Publications (1959)], and applications to industrial problems are given by Cheremisinoff (1979, 1981a,b).

The first of these devices, the *pitot tube,* is used for measuring local fluid velocities. The device consists of a stainless steel tube with its inlet opening turned upstream into the flow. Therefore, the inlet receives the full impact of the flow against it. The impact is converted completely into pressure head $w^2/2g$, superimposed on the existing static pressure of the fluid. In principle, the pitot tube consists of both an impact tube and a piezometer tube [Figure 5.14(a)]. Because an *impact tube* is used in connection with a *piezometer tube,* the static pressure may be subtracted from the total pressure measured, the difference of which is the velocity head. This has been illustrated in Figure 5.12. The pressure difference is measured conveniently with a differential U-tube, as shown in Figure 5.14(a). Note that the U-tube should contain a liquid that does not mix with the working fluid and has higher density. Both the impact and piezometer tubes are combined into a standard S-shaped pitot tube by including static pressure taps downstream of the impact tube's tip. The design is illustrated in Figure 5.14(b).

The maximum flow velocity along a pipe axis may be computed from the measured pressure head $w^2/2g$, using the Bernoulli equation. Consider, for example, an incompressible fluid flowing between points 1 and 2 in Figure 5.14(a). As the velocity at 2 is zero, we may write

$$\frac{w_1^2}{2g} = \frac{P_2 - P_1}{\varrho} \qquad (5.126)$$

or

$$w = C\sqrt{\frac{2g\Delta p}{\varrho}} \qquad (5.127)$$

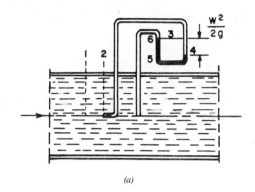

(a)

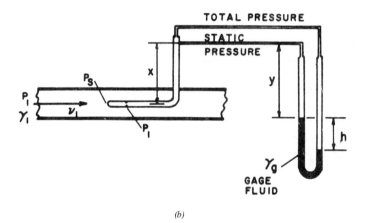

(b)

FIGURE 5.14. Pitot tube used to measure the maximum fluid velocity in a pipe.

where C is the pitot coefficient obtained from calibration. For a well-designed pitot tube, C has a value between 0.96 and 0.98.

The value of Δp may be obtained from the Bernoulli equation for the case of zero flow:

$$\Delta p = -\varrho \Delta z g \qquad (5.128)$$

From Figure 5.14(a), we see that

$$\Delta p = P_2 - P_1 = (P_2 - P_3) + (P_3 - P_4) + (P_4 - P_5)$$
$$= (P_5 - P_6) + (P_6 - P_1) \qquad (5.129)$$

As $P_4 - P_5 = 0$ and $P_2 - P_3 = P_6 - P_1$, Equation (5.129) may be simplified to

$$\Delta p = (P_3 - P_4) + (P_5 - P_6) \tag{5.130}$$

from Equation (5.128) and noting that $z_3 - z_4 = z_5 - z_6$, we obtain

$$\Delta p = g(\varrho_m - \varrho)(z_3 - z_4) \tag{5.131}$$

Combining Equations (5.127) and (5.131),

$$w = C\sqrt{\frac{2g(\varrho_m - \varrho)\Delta h}{\varrho}} \tag{5.132}$$

where ϱ_m is the density of the manometer liquid and $\Delta h = z_3 - z_4$.

To determine the average fluid velocity, measurements at successive points across the pipe cross section are needed. The velocity profile obtained from traversing the pipe may then be integrated over the pipe diameter to obtain an average value. Such a method for measuring velocity and flow rate is simple; however, in general it is not accurate due to the difficulties in positioning the instrument exactly along the pipe axis.

Further discussions on the pitot tube are given by Cheremisinoff (1979), the ASME Research Committee on Fluid Meters (1959), Stoll (1959), and Folsom (1956).

Example 5.11

For the pitot tube arrangement shown in Figure 5.14(b), water is flowing through the pipe at 130°F. The manometer fluid is mercury ($\gamma = 13.6$). The manometer reading is 15 in, and the pitot tube coefficient is 0.97. Determine the point velocity of the water.

Solution

Equation (5.132) will be used:

$$w_1 = C\sqrt{2g(\varrho_m - \varrho)\Delta h/p}$$

$$\varrho = 61.2 \text{ lb}_m/\text{ft}^3$$

$$\varrho_m = (13.60)(62.4 \text{ lb}_m/\text{ft}^3) = 848 \text{ lb}_m/\text{ft}^3$$

$$\Delta h = (15 \text{ in})(1 \text{ ft}/12 \text{ in}) = 1.25 \text{ ft}$$

The velocity in English units is

$$w_1 = 0.97 \sqrt{\frac{(2)(32.2)(848 - 61.2)(1.25)}{61.2}}$$

$$= 31.2 \text{ ft/s}$$

An *orifice meter* consists of a thin plate mounted between two flanges, with an accurately drilled hole positioned concentric to the pipe axis. The flow measurement principle behind the orifice meter is based on the reduction of flow pressure accompanied by an increase in velocity, i.e., Bernoulli's principle. The reduction of the cross section of the flowing stream as it passes through the orifice increases the velocity head at the expense of pressure head. Figure 5.15 illustrates the operation of an orifice meter whereby a *manometer* or pressure gauge is used to measure the upstream and downstream pressures. By applying the Bernoulli equation, the discharge can be determined from the manometer's readings based on the known area of the orifice.

A *venturi meter,* illustrated in Figure 5.16, consists of a short length of straight tubing connected at either end of the pipe by conical sections. The measurement principle behind this instrument is based on the reduction of flow pressure accompanied by an increase in velocity of the venturi throat. The pressure drop experienced in the upstream cone section is used to measure the rate of flow through the venturi meter. On the discharge side of the meter the fluid velocity is decreased and the original pressure recovered. Because of its shape, pressure losses in a venturi meter are less than in an orifice meter. However, a venturi meter is large in comparison to an orifice meter, which can be mounted readily between flanges. [Detail design and performance data of venturis are given by Cheremisinoff (1979), ASME Committee on Fluid Meters (1959), Miner (1956), and Hooper (1950)].

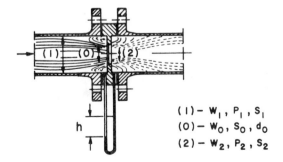

FIGURE 5.15. An orifice meter in operation.

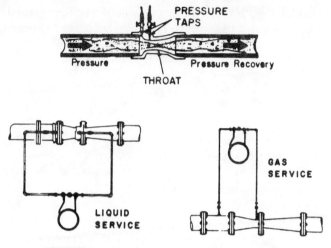

FIGURE 5.16. Flow measurement using a venturi meter.

Because of its size and expense, a smaller version of the venturi, called the *flow nozzle*, has been developed. A typical flow nozzle is illustrated in Figure 5.17. It is often used as the primary element for measuring liquid flows. In the design, the diverging exit of the venturi meter is omitted and the converging entrance altered to a more rounded configuration. In both the venturi meter and flow nozzle, the cross-sectional area of the compressed stream ($S_2 = \pi d_1^2/4$) is equal to the cross-sectional area of the hole $S_o = \pi d^2/4$. In the immediate discussions to follow, we shall derive formulas applicable to orifice meters, venturi meters, and flow nozzles.

Let us apply the Bernoulli equation over two sections of horizontal pipe

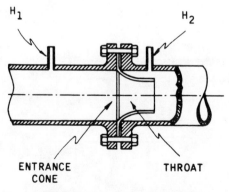

FIGURE 5.17. A typical flow nozzle.

flow. The pressure drop between these two sections may be measured with a differential U-tube. Using the notation given in Figure 5.15, we obtain

$$\frac{P_1}{\varrho g} + \frac{w_1^2}{2g} = \frac{P_2}{\varrho g} + \frac{w_2^2}{2g}$$

Hence,

$$\frac{w_2^2}{2g} - \frac{w_1^2}{2g} = \frac{P_1 - P_2}{\varrho g} = \Delta h \qquad (5.133)$$

where Δh is the pressure drop measured by the differential U-tube in m of working liquid column. Using the continuity equation, we determine the average velocity and fluid rate in the piping. w_1 denotes the average velocity in the pipe's large cross section and refers to the stream flow immediately after the orifice where pressure p_2 is measured:

$$w_1 = w_2 \frac{S_2}{S_1} = w_2 \frac{d_2^2}{d_1^2} \qquad (5.134)$$

Substituting the value w_1 from Equation (5.134) into Equation (5.133), we obtain

$$\frac{w_2^2}{2g} - \frac{w_2^2}{2g}\left(\frac{d_2}{d_1}\right)^4 = \Delta h \qquad (5.135)$$

Hence,

$$w_2 = \sqrt{\frac{2g\Delta h}{1 - \left(\dfrac{d_2}{d_1}\right)^4}} \qquad (5.136)$$

The volume liquid rate V_{sec} in section S_0 of the hole in the orifice meter (and, consequently, in the piping) is

$$V_{sec} = \frac{\alpha \pi}{4} d_0^2 \sqrt{\frac{2g\Delta h}{1 - \left(\dfrac{d_0}{d_1}\right)^4}} \qquad (5.137)$$

where α is the discharge coefficient $\alpha = f(Re\ d_0/d_1)$.

The term (d_2/d_1) in the denominator of Equation (5.137) is usually small. Hence, as a first approximation the volumetric flow rate may be computed from

$$V_{sec} = \frac{\alpha \pi}{4} d_0^2 \sqrt{2g\Delta h} \tag{5.138}$$

Also, the mean velocity through the pipe may be approximated by

$$\overline{W} = \alpha \left(\frac{d_0}{d}\right)^2 \sqrt{2g\Delta h} \tag{5.139}$$

For compressible fluids, a correction coefficient must be applied to Equations (5.138) and (5.139). The following examples illustrate the use of the above formulas.

Example 5.12

A liquid of density 1237 kg/m³ and viscosity 0.74 cp is flowing through a 19-cm i.d. pipe. A sharp-edged orifice having a diameter of 2.9 cm is installed in the pipeline. The measured pressure drop across the orifice is 112.7 kN/m². Calculate the volumetric flow rate and the average velocity of the liquid through the pipe.

Solution

Equation (5.137) is used:

$$V_{sec} = \frac{\alpha \pi}{4} d_0^2 \sqrt{\frac{2(P_1 - P_2)/\varrho}{1 - (d_0/d_1)^4}}$$

$$P_1 - P_2 = 112.7 \text{ kN/m}^2 = 11.27 \times 10^4 \text{ N/m}^2$$

$$d_1 = 0.190 \text{ m} \qquad d_0 = 0.029 \text{ m} \qquad \frac{d_0}{d_1} = \frac{0.029}{0.190} = 0.153$$

Examining the discharge coefficient versus the Reynolds number plot (Figure 5.18), we note that for $Re > 20{,}000$, α is roughly the same regardless of the diameter ratio. Hence, we shall assume that $\alpha \cong 0.61$:

$$V_{sec} = \frac{(0.61)\pi}{4} (0.029)^2 \sqrt{\frac{2(11.27 \times 10^4)/1237}{1 - (0.153)^4}}$$

$$V_{sec} = 0.00544 \text{ m}^3/\text{s (or 1.44 gpm)}$$

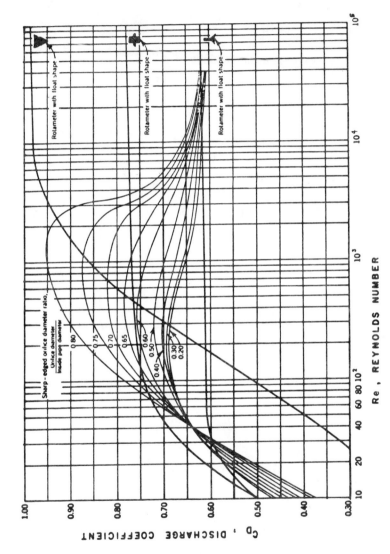

FIGURE 5.18. Plot of discharge coefficient versus Reynolds number for sharp-edged orifices and rotameters [from Brown et al. (1950)].

The average velocity of the liquid is

$$\overline{W} = \frac{0.00544 \text{ m}^3/\text{s}}{\frac{\pi}{4}(0.190)^2 \text{ m}^2} = 0.192 \text{ m/s (or 0.63 fps)}$$

Re is calculated to determine whether it is greater than 20,000 for $\alpha = 0.61$. $\mu = 0.74 \times 10^{-3}$ kg/m-s $= 0.74 \times 10^{-3}$ Pa-s:

$$Re = \frac{d_1 W \varrho}{\mu} = \frac{(0.190)(0.192)(1237)}{0.74 \times 10^{-3}} = 60,981$$

As the Reynolds number is greater than 20,000, a good value for the discharge coefficient was selected. The student should repeat the problem using Equation (5.139) and compare the results to the above.

Example 5.13

The venturi meter shown in Figure 5.16 is used to measure the flow rate in a pipe. The upstream diameter of the venturi is 4.5 in (d_1) and the throat diameter $(d_0 = d_2)$ is 1.75 in. The pressure drop across the meter is 0.290 psi and the fluid is air $(\varrho \cong 0.07 \text{ lb}_m/\text{ft}^3)$.

Solution

From continuity,

$$V = F_1 w_1 = F_2 w_2$$

where V is the volumetric discharge. Thus,

$$\frac{\pi}{4}\left(\frac{4.5}{12}\right)^2 w_1 = \frac{\pi}{4}\left(\frac{1.75}{12}\right)^2 w_2$$

or

$$V = 0.1104 \, w_1 = 0.0167 \, w_2$$

Applying the Bernoulli equation for $z_1 = z_2$,

$$P_1 - P_2 = 0.290 \times 144 = 41.76 \text{ lb}_f/\text{ft}^2$$

$$\frac{P_1 - P_2}{\varrho} = \frac{w_2^2}{2g} - \frac{w_1^2}{2g}$$

$$\frac{41.76}{0.07} = \frac{V^2}{2g}\left[\left(\frac{1}{0.0167}\right)^2 - \left(\frac{1}{0.1104}\right)^2\right]$$

Solving for the volumetric discharge gives $V = 10.97$ cfs (or 2764 lb_m/hr of air flowing).

Efflux from Vessels and Pipes

We now direct our attention to another common application of Bernoulli's principle, that of efflux from a tank. Consider the system shown in Figure 5.19(a), in which water is being discharged through a circular orifice located on the floor of a flat bottomed vessel open to the atmosphere. The tank has an inlet stream that maintains a constant liquid level H. The reference plane 0-0 is designated to be below and parallel to both, plane 1-1 corresponding to the liquid surface and plane 2-2 passing through the narrowest section of discharging jet. Hence, for an ideal liquid, Bernoulli's equation is

$$z_1 + \frac{P_1}{\varrho g} + \frac{w_1^2}{2g} = z_2 + \frac{P_2}{\varrho g} + \frac{w_2^2}{2g}$$

As this is an open vessel with constant level, $P_1 = P_2$ and $w_1 = 0$. Neglecting the small distance between the orifice plane at the vessel floor and plane

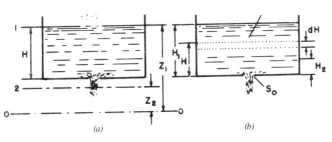

FIGURE 5.19. Efflux from a tank: (a) at constant liquid level in the tank; (b) at variable liquid level in the tank.

2-2, we may assume that $z_1 - z_2 = H$. Hence,

$$\frac{w_2^2}{2g} = H$$

and

$$w_2 = \sqrt{2gH}$$

For viscous fluids, a portion of the liquid heat is lost to friction and to overcoming the resistance resulting from the abrupt jet constriction in the orifice. Therefore, the discharge or jet velocity of a viscous liquid is the following:

$$w_2 = \psi\sqrt{2gH}$$

where ψ is a correction factor called the velocity factor which accounts for head losses incurred by discharging through the orifice. Values for ψ are less than 1. Because the jet cross-sectional area in the orifice S_0 is larger than its narrowest section S_2, the liquid velocity w_0 in the orifice must be less than w_2. Then

$$w_0 = \epsilon_j w_2 = \epsilon_j \psi \sqrt{2gH} = \alpha\sqrt{2gH} \qquad (5.140)$$

where $\epsilon_j = S_2/S_0$ is the jet constriction coefficient. Coefficient α is the discharge coefficient defined as the product of the velocity coefficient ψ and the constriction coefficient ϵ_j:

$$\alpha = \psi\epsilon_j \qquad (5.141)$$

Values of discharge coefficient α may be found in handbooks such as Perry's (1950). The coefficient is a function of the Reynolds number and is determined experimentally.

The volume liquid rate is equal to the velocity in the orifice w_0 times the orifice cross-sectional area S_0:

$$V_{sec} = \alpha S_0 \sqrt{2gH} \qquad (5.142)$$

Equation (5.142) shows that the rate of efflux through the orifice depends on the constant level in the tank and the orifice diameter but is independent of the vessel's shape. The expression is also valid for liquid discharging through an orifice in the tank's side wall. For the latter case, the level H must be measured from the centerline of the orifice.

For liquids in which viscosities are similar to that of water, $\alpha = 0.62$.

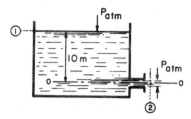

FIGURE 5.20. System for tank discharge problem (Example 5.14).

Example 5.14

Water is being discharged through a side wall nozzle in a large tank open to the atmosphere (Figure 5.20). The jet issues as a cylinder to the atmosphere and the water surface is located 10 m above the nozzle centerline. Determine the velocity of efflux from the nozzle.

Solution

This involves application of Bernoulli's principle:

$$z_1 + \frac{P_2}{\gamma_1} + \frac{w_1^2}{2g} = z_2 + \frac{P_2}{\gamma_2} + \frac{w_2^2}{2g}$$

where subscript 2 refers to a point immediately downstream of the nozzle (along its centerline) and 1 refers to the level of the tank. We may assume that the pressure along the centerline of the nozzle discharge is atmospheric. Hence, $P_1 - P_2 = 0$. And, as the elevation datum is at point 2, $z_2 = 0$ and $z_1 = H = 10$ m.

Because this is a large vessel, we may further assume that the velocity of the reservoir is practically zero. Hence,

$$0 + 0 + H = \frac{w_2^2}{2g} + 0 + 0$$

or

$$w_2 = \sqrt{2gH} \qquad \text{(derived earlier)}$$

$$w_2 = \sqrt{2 \times 9.806 \times 10} = 14.0 \text{ m/s}$$

Our solution has ignored frictional effects due to the nozzle; therefore, it is a maximum. It states that the velocity of efflux is equal to the velocity of free fall from the surface of the tank. This is known as *Torricelli's theorem*.

The student should continue with the problem by calculating w_2 for $\psi = 0.75$. For a fluid jet diameter of 55 mm, determine the volumetric discharge.

In short cylindrical nozzles, additional head losses in both the nozzle inlet and outlet make frictional corrections even more important (that is, ψ becomes smaller). However, the fluid entering a nozzle tends to expand after some constriction, eventually filling the entire nozzle cross section, i.e., it can be assumed that $\epsilon_j = 1$. The end result is that nozzles generally have a higher discharge coefficient than a simple orifice (for water discharge, $\alpha \cong 0.82$).

Let us now consider the second efflux problem shown in Figure 5.19(b), that of tank discharge from a variable-level tank. Specifically, we wish to determine the discharge time through an orifice located on the floor of a thin-walled vessel. For a time interval dt the liquid level H_1 is decreased to some height H_2. According to Equation (5.142), the liquid volume discharged from the vessel is

$$dV = V_{sec} dt = \alpha S_0 \sqrt{2gH} \, dt \qquad (5.143\text{A})$$

where S_0 is the cross-sectional area in the bottom of the vessel.

Over the same time interval dt, the liquid level is lowered by the differential height dH, and the differential volume lost for a vessel of constant cross-sectional area S is

$$dV = -SdH \qquad (5.143\text{B})$$

From the principle of continuity, Equations (5.143A) and (5.143B) may be equated to each other:

$$\alpha S_0 \sqrt{2gH} \, dt = -SdH$$

Hence,

$$dt = \frac{-SdH}{\alpha S_0 \sqrt{2gH}} \qquad (5.144)$$

Assuming constant α and integrating over the limits of H_1 to H_2 for 0 to t, an expression for the time of discharge for a vessel of constant cross section is obtained:

$$\int_0^t dt = \int_{H_1}^{H_2} \frac{SdH}{\alpha S_0 \sqrt{2gH}}$$

$$\qquad (5.145)$$

$$t = \frac{S}{\alpha S_0 \sqrt{2g}} \int_{H_1}^{H_2} H^{-1/2} dH = \frac{2S}{\alpha S_0 \sqrt{2g}} (\sqrt{H_1} - \sqrt{H_2})$$

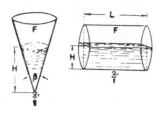

FIGURE 5.21. Liquid discharge from vessels of variable cross section.

If the vessel is emptied completely, i.e., $H_2 = 0$, the expression simplifies to the following:

$$t = \frac{2S\sqrt{H_1}}{\alpha S_0 \sqrt{2g}} \tag{5.146}$$

There are many operations in which vessels of variable cross sections are utilized (for example, conical vessels, horizontal cisterns, etc.). To determine the discharge times from such vessels, the relationship between cross-sectional area and height must be known, i.e., $S = f(H)$. Consider the two systems illustrated in Figure 5.21: a funnel-shaped vessel and a horizontal cylindrical tank. The conical vessel with apex angle β has a cross-sectional area equal to the following:

$$S = \pi H^2 \tan^2\beta/2 \tag{5.147}$$

The horizontal cylindrical vessel's cross section is

$$S = 2L\sqrt{HD - H^2} \tag{5.148}$$

When these relationships are included in Equation (5.145), the discharge time between any two levels or for complete discharge can be computed for the two geometries.

Another common system encountered in plant operations is liquid discharge from partially filled pipes. Such a system is shown in Figure 5.22. The volumetric discharge rate depends on the pipe diameter D and a space factor

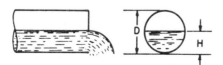

FIGURE 5.22. Liquid discharge from a partially filled pipe.

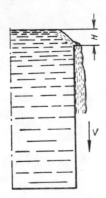

FIGURE 5.23. Flow over a weir.

defined by $K = H/D$. For water, the discharge rate may be obtained from the following empirical formula [refer to Folsom (1956) for details].

$$V = 2.54 D^{2.56} K^{1.84} \qquad (5.149)$$

where V is in units of ℓ/min and D is in m. Equation (5.149) is valid for $K = 0.2$–0.6 and for $D = 3$–15 cm.

Figure 5.23 illustrates still another discharge system, that of flow over a weir. For weir flow, it is important to know the crest over a weir for a given perimeter L to determine the overflow rate. Whitwell and Plumb (1939) developed the following empirical formula for water flow:

$$\frac{V}{L} = 304(H - 0.0032) \qquad (5.150)$$

where V has units of ℓ/s and L and H are in m. Equation (5.150) is applicable for $H < 0.001$ m. For $0.01 < H < 0.3$ m, the following formula is applicable:

$$\frac{V}{L} = 1670 H^{1.455} \qquad (5.151)$$

The above weir formulas were obtained with sharp-crested weirs. The shape of the weir edge (i.e., sharp or blunted) generally does not have a significant effect on the discharge rate.

Example 5.15

Determine the discharge time for a horizontal cylindrical cistern having flat ends 2 m in diameter and an overall length of 4 m. The cistern is filled with

12 tons of a liquid whose specific gravity is 1.1 and has a **viscosity** close to water. The diameter of the discharge hole is 27 mm.

Solution

The discharge time may be computed from Equation (5.145). We shall assume $\alpha = 0.62$. The cross-sectional area of the discharge opening is

$$S_o = \frac{3.14 \times 0.027^2}{4} = 5.7 \times 10^{-4}$$

Hence,

$$t = \frac{1}{0.62 \times 5.7 \times 10^{-4} \sqrt{2 \times 9.81}} \int_0^{H_1} SH^{-1/2} dH = 637 \int \frac{F}{\sqrt{H}} dH, \text{ s}$$

where $F = L \times S$ (Figure 5.21).

We can write the following relationship for the tank radius:

$$R^2 = \left(\frac{S}{2}\right)^2 + (H - R)^2$$

where $R = D/2$.

Solving for the tank's cross section,

$$S = 2\sqrt{R^2 - (H - R)^2} = 2\sqrt{H(2R - H)} = 2\sqrt{H(D - H)}$$

Hence,

$$F = 2L\sqrt{H(D - H)} = 8\sqrt{H(2 - H)}$$

Substituting back into our expression,

$$t = 637 \int_0^{H_1} \frac{8H(2 - H)}{\sqrt{H}} dH = 5.1 \times 10^3 \int_0^{H_1} \sqrt{2 - H} \, dH$$

or

$$t = 9.62 \times 10^3 - 3.4 \times 10^3 (2 - H_1)^{3/2}, \text{ s}$$

H_1 is the initial liquid height, which is unknown. However,

$$V = \frac{12,000}{1,100} = 10.9 \text{ m}^3 \text{ and } L = 4 \text{ m}$$

Therefore, $F_1 = 10.9/4 = 2.72$ m², where F_1 is the initial vertical section of liquid.

The vertical section of the cistern is

$$F_2 = \pi D^2/4 = 3.14 \times 2^2/4 = 3.14 \text{ m}^2$$

Hence, the cistern vertical free section is

$$F_3 = 3.14 - 2.72 = 0.42 \text{ m}^2$$

For the ratio $F_3/D^2 = 0.105$, we find in Perry's *Chemical Engineer's Handbook* (1950) $H/D = 0.1915$. Therefore, $H = 0.1915 \times 2 = 0.383$ m, and the initial liquid depth in the cistern is

$$H_1 = D - H = 2 - 0.383 = 1.62 \text{ m}$$

Finally, discharge time, $t = 9620 - 3400(2 - 1.62)^{3/2} = 8824$ s, or

$$t = \frac{8824}{3600} = 2.5 \text{ hr}$$

HYDRAULIC RESISTANCES IN PIPE FLOW

Flow through Pipes

A large chemical plant or refinery is analogous to the human body. Pumps and compressors perform functions similar to the heart, moving various chemicals and fluids to different reactors, which are the vital organs of the plant. Piping networks and ducts are the veins and arteries that transport the plant's processing fluids. The makeup of an entire plant can be almost as complex as the human body. As with most healthy humans, nature already has applied rigorous principles of hydrodynamics to the proper size of our hearts, to the energy consumption required by the heart, and to the most appropriate layout of the arterial system.

In a similar manner, although perhaps less rigorously, we apply the same hydrodynamic principles to the distribution systems in plant operations. We direct our attention now to the determination of head losses or pressure drops. By determining such hydraulic resistances in piping dstributions, proper selection and sizing of the plant's heart elements can be made, namely, the required energy ratings of pumps, compressors, fans, etc.

For the most general case, head losses are evaluated from friction and local resistances. *Frictional resistance* arises from the motion of real fluids through piping, and its value is influenced by the flow regime. Thus, in turbulent flow, resistances are characterized not only by viscosity but by the eddy viscosity, which depends on the hydrodynamic conditions. Such resistances result in additional energy losses.

Local resistances arise due to variations in fluid velocity caused by changes in flow area or direction. Among the many local resistances are friction losses caused by entrances and exits from piping, sudden contractions and expansions, losses in fittings, valves, etc. Therefore, the loss of head h_ℓ may be expressed by the sum of two terms:

$$h_\ell = h_{fr} + h_{\ell r} \qquad (5.152)$$

where h_{fr} and $h_{\ell r}$ are head losses due to friction and local resistances, respectively.

Friction losses for a laminar flow in a straight pipe may be determined analytically from the Poiseuille equation [Equation (5.27B)].

For a horizontal pipe of constant cross section, i.e., $z_1 = z_2$ and $w_1 = w_2$, Bernoulli's equation is applied to determine the head loss due to friction:

$$\frac{P_1 - P_2}{\varrho g} = \frac{\Delta P}{\varrho g} = h_{fr}$$

Substituting $\Delta P = \varrho g h_{fr}$ into Equation (5.27B) and noting that V_{sec} may be estimated by the product of average velocity and the pipe's cross-sectional area $\pi d^2/4$, we obtain

$$\overline{W} = \frac{\pi d 2}{4} = \frac{\pi d^4 \varrho g h_{fr}}{128 \mu \ell}$$

where ℓ and d are the length and diameter of the pipe. By simplifying and solving for the head loss, we arrive at

$$h_{fr} = \frac{32 \overline{W} \mu \ell}{\varrho g d^2}$$

Multiplying the numerator and denominator of the RHS of this expression by $2\overline{W}$, we obtain

$$h_{fr} = \frac{64\mu}{\overline{W}d\varrho} \frac{\ell}{\delta} \frac{\overline{W}^2}{2g}$$

or

$$h_{fr} = \frac{64}{Re} \frac{\ell}{d} \frac{\overline{W}^2}{2g} \qquad (5.153)$$

Thus, the head losses due to friction are expressed in terms of the velocity head, $\overline{W}^2/2g$. The quantity $64/Re$ in Equation (5.153) is referred to as the *friction factor* for laminar flow, which we shall denote by λ.

Further, we denote ζ as the friction resistance coefficient defined as follows:

$$\zeta = \lambda \frac{\ell}{d} \qquad (5.154)$$

Using this notation, Equation (5.153) is expressed in the following form:

$$h_{fr} = \zeta \frac{\overline{W}^2}{2g} = \lambda \frac{\ell}{d} \frac{\overline{W}^2}{2g} \qquad (5.155)$$

And, for pressure losses ($\Delta P_{fr} = \varrho g h_{fr}$),

$$\Delta P_{fr} = \lambda \frac{\ell}{d} \frac{\varrho \overline{W}^2}{2} \qquad (5.156)$$

Equation (5.155) has been found to agree well with experimental data for laminar flow, i.e., $Re < 2100$. For laminar flow, the friction factor is practically independent of pipe roughness.

Example 5.16

A small capillary tube with an inside diameter of 2.9 mm and length of 0.5 m is being used to measure the viscosity of a liquid having a density of 920 kg/m³. The pressure drop reading across the capillary during flow is 0.08 m (water density = 998 kg/m³). The volumetric flow rate through the capillary is 9.27×10^{-7} m³/s. Ignoring end effects, determine the viscosity of the liquid.

Solution

First convert height h of 0.08 m of water to a pressure:

$$\Delta p = \varrho g h_{fr}$$

$$= \left(998 \frac{kg}{m^3}\right)\left(9.80665 \frac{m}{s^2}\right)(0.08 \text{ m})$$

$$= 783 \text{ kg-m/s}^2\text{-m}^2 = 783 \text{ N/m}^2 \text{ (or 0.114 psia)}$$

Assuming the flow to be laminar, we apply Poiseuille's equation [Equation (5.27B)], solving for viscosity μ:

$$\mu = \frac{\pi d^4 \Delta p_f}{128 \ell V_{sec}}$$

where

$\ell = 0.5$ m
$d = 2.9 \times 10^{-3}$ m
$V_{sec} = 9.27 \times 10^{-7}$ m³/s
$\Delta p_f = 783$ Nm²

Substituting in values,

$$\mu = \frac{\pi (2.9 \times 10^{-3} \text{ m})^4 (783 \text{ N/m}^2)}{128(0.5 \text{ m})\left(9.27 \times 10^{-7} \frac{m^3}{s}\right)}$$

$$= 2.933 \times 10^{-3} \frac{kg}{m\text{-}s} = 2.933 \times 10^{-2} \text{ p (or 2.93 cp)}$$

Since it was assumed that the flow is laminar, the Reynolds number should be computed as a check:

$$\overline{W} = \frac{4 V_{sec}}{\pi d^2} = \frac{4\left(9.27 \times 10^{-7} \frac{m^3}{s}\right)}{\pi (2.9 \times 10^{-3} \text{ m})^2} = 0.1403 \frac{m}{s}$$

$$Re = \frac{d \overline{W} \varrho}{\mu} = \frac{(2.9 \times 10^{-3} \text{ m})\left(0.1403 \frac{m}{s}\right)\left(920 \frac{kg}{m^3}\right)}{2.933 \times 10^{-3} \frac{kg}{m\text{-}s}} = 128$$

Thus, the flow is laminar as assumed.

Example 5.17

For the previous example problem, assume that the viscosity is known and the pressure drop Δp_f is to be predicted. Using the friction factor expression for laminar flow [Equation (5.156)], compute Δp_f and compare.

Solution

The Reynolds number computed for Example 5.16 is $Re = 128$. Hence, the friction factor is

$$\lambda = \frac{64}{Re} = \frac{64}{128} = 0.50$$

Using Equation (5.156),

$$\Delta p_f = \lambda \left(\frac{\ell}{d}\right)\left(\frac{\varrho \overline{W}^2}{2}\right) = 0.50 \left(\frac{0.5 \text{ m}}{2.9 \times 10^{-3} \text{ m}}\right)\left(\frac{920 \frac{\text{kg}}{\text{m}^3}}{2}\right)\left(0.1403 \frac{\text{m}}{\text{s}}\right)^2$$

$$= 781 \frac{\text{kg}}{\text{m}^2 \cdot \text{s}}$$

This agrees very well with the previous problem. Additional illustrative examples are given by Dodge (1968), King and Brater (1963), and King and Crocker (1967).

Equation (5.155) may also be used for conduits of noncircular cross sections, where diameter d is replaced by an equivalent diameter. The equivalent diameter d_{eq} is defined as 4 times the hydraulic radius, where the hydraulic radius is the cross-sectional area of flow divided by the wetted perimeter. In applying Equation (5.155), the laminar friction factor expression changes to the following:

$$\lambda = \frac{B}{Re} \qquad (5.157)$$

where $B = 57$ for a square cross section and $B = 96$ for an annular cross section. Equation (5.155) may also be applied to determine friction losses in the turbulent regime [see Hodson (1939), Rouse (1948), and Davies (1972)]. However, in this case the friction factor relationship cannot be derived theoretically. Therefore, design equations for estimating λ in turbulent flow are obtained by generalized experimental data through the use of similarity theory. The generalized expression may be transformed into an exponential form:

$$Eu = A\, Re^m \left(\frac{\ell}{d}\right)^q \qquad (5.158\text{A})$$

From normalized data for fluid motion through smooth-walled piping in the Reynolds number range of 4000 to 100,000, the following constants in Equation (5.158) were evaluated:

$$A = 0.158, \qquad m = -0.25, \qquad q = 1$$

Hence, the design expression is

$$Eu = 0.158\, Re^{-0.25} \left(\frac{\ell}{d}\right) \qquad (5.158\text{B})$$

or

$$h_{fr} = 0.316\, \frac{\ell}{d}\, \frac{\overline{W}^2}{2g} \qquad (5.158\text{C})$$

Comparing Equations (5.158C) and (5.155) reveals that the friction factor may be expressed in the following form:

$$\lambda = 0.316\, Re^{-0.25} \qquad (5.159)$$

Hence, for laminar flow the frictional pressure drop is proportional to fluid velocity raised to the first power [Equation (5.153)], whereas for the turbulent regime it is proportional to velocity raised to the 1.75 power.

The friction factor depends not only on the Reynolds number, but also on the roughness of the pipe walls. Roughness is characterized by the effective height of the protrusions, e.g., Figure 5.24 shows a plot of friction factor versus Reynolds number for different piping materials. As noted earlier, the laminar friction factor shows no dependency on the condition of the pipe. However, a significant difference in the λ–Re relationship is seen for different material pipes in the turbulent regime. As shown by curve 2, rough pipes tend to lower λ, which increases frictional resistance. To normalize the effect of roughness, Moody (1944) developed the generalized friction factor plot shown in Figure 5.25. The ordinate on the right of Figure 5.25 is the inverse of relative roughness, where $\epsilon = e/d$.

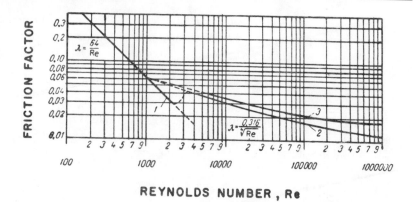

FIGURE 5.24. Friction factor vs Reynolds number chart: curve 1 = laminar flow for smooth and rough pipes; curve 2 = turbulent flow for smooth pipes made from copper, lead, and glass; curve 3 = turbulent flow for rough pipes made from steel and cast iron.

In laminar flow through a pipe in which e/d is usually somewhat less than 0.01, the influence of wall roughness is insignificant, because the fluid fills the spaces between the protrusions, and the inner fluid layers slide smoothly over a pipe of effective diameter $d - 2e$. In some initial range of turbulence, the wall roughness can also be neglected if it is smaller in height than the thickness of the viscous sublayer. In this case, the pipe is said to be hydraulically smooth; λ then can be calculated using Equation (5.159).

As the Reynolds number increases, the thickness of the viscous sublayer decreases. When the viscous sublayer becomes comparable to the effective height of wall protrusions, disturbances enter into the bulk flow, thus increasing turbulence and flow resistance. Under these conditions, the friction factor becomes more strongly dependent on roughness. Thus, λ and frictional pressure drop increase under the action of inertia forces due to the additional formation of vortexes about the protrusions.

From the above, it becomes apparent that as the Reynolds number increases, three regimes of flow or zones develop. The first is referred to as the zone of *smooth friction*, where $\lambda = f(Re)$. The second is the zone of *mixed friction*, where λ is both a function of Re and roughness. Finally, there is the *self-modeling* zone, where λ becomes practically independent of the Reynolds number and is primarily a function of wall roughness. The self-modeling zone is also called the *zone of quadratic resistance law* as $\lambda = f(\overline{W}^2)$.

The critical Reynolds numbers $Re_{cr,1}$, where roughness begins to influence λ, as well as critical values $Re_{cr,2}$, when λ is only a function of pipe roughness (refer to dashed line in Figure 5.25), depend on the *relative*

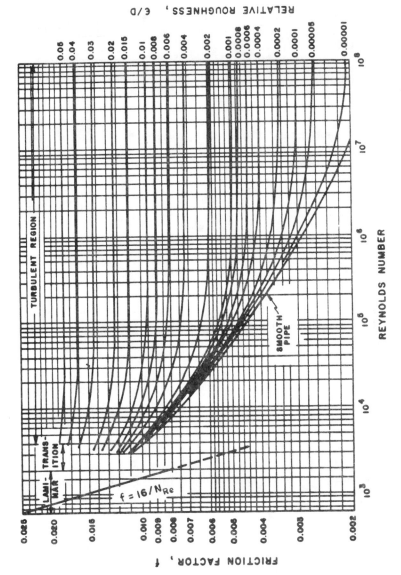

FIGURE 5.25. Reynolds number–friction factor chart as a function of relative roughness.

roughness. These critical values have been determined experimentally to be the following:

$$Re_{cr,1} \cong \frac{23}{\epsilon} \qquad (5.160)$$

and

$$Re_{cr,2} \cong 220\epsilon^{-9/8} \qquad (5.161)$$

Hodson (1939) recommends the following correlation for all three zones of turbulent motion:

$$\frac{1}{\sqrt{\lambda}} = -2 \log\left[\frac{\epsilon}{3.7} + \left(\frac{6.81}{Re}\right)^{0.9}\right] \qquad (5.162)$$

For the zone of smooth friction, λ can be calculated either from Equation (5.159) or from Equation (5.162). In the latter equation, the first term inside the parentheses is eliminated. This term reflects the influence of roughness, which, for this zone, is insignificant. Hence,

$$\frac{1}{\sqrt{\lambda}} = -2 \log\left(\frac{6.81}{Re}\right)^{0.9} = 1.8 \log Re - 1.5 \qquad (5.163)$$

For the self-modeling zone, where λ is independent of Re, Equation (5.163) has the following form:

$$\frac{1}{\sqrt{\lambda}} = 2 \log \frac{3.7}{\epsilon} \qquad (5.164)$$

Further clarification of the definition of the friction factor is needed at this point. In this text we have elected to adopt the friction factor originally defined by Blasius and later used by Moody (1944) and others. The Blasius friction factor is defined as four times the Fanning friction factor (so, for laminar flow, $\lambda_F = 16/Re$). The chart given in Figure 5.25 is based on the Blasius friction factor. Care must be exercised when performing calculations of frictional losses as some authors prefer to define a friction factor that is twice the size of the Fanning friction factor.

The following examples apply friction factors to estimate friction loss for steady flow in uniform circular pipes running full of liquid under isothermal conditions.

Example 5.18

Water is flowing full through a 50.8-mm (2.0-in.) i.d. wrought iron pipe at a rate of 2.7×10^{-3} m³/s. The viscosity and density of the water under the process conditions are 1.09 cp and 992 kg/m³, respectively. Determine the mechanical energy friction loss for a 73-m section of pipe.

Solution

The following information is known:
- $d = 0.0508$ m
- $\varrho = 992$ kg/m³
- $\ell = 73$ m
- $\mu = (1.09 \text{ cp})(1 \times 10^{-3}) = 1.09 \times 10^{-3}$ kg/m-s

The cross-sectional area of flow is

$$F = \frac{\pi}{4}(0.0508 \text{ m})^2 = 2.027 \times 10^{-3} \text{ m}^2$$

Hence, the liquid velocity is

$$w = \frac{2.7 \times 10^{-3} \text{ m}^3/\text{s}}{2.027 \times 10^{-3} \text{ m}^2} = 1.33 \text{ m/s}$$

The Reynolds number is computed as

$$Re = \frac{dw\varrho}{\mu} = \frac{(0.0508)(1.33)(992)}{1.09 \times 10^{-3}} = 61{,}588$$

Hence, the flow is turbulent. For wrought iron pipe, a typical value for surface roughness is 0.00015 ft. Hence, the equivalent roughness is

$$\frac{e}{d} = \frac{0.00015 \text{ ft}}{0.0508 \text{ m}} \times \frac{\text{m}}{3.28 \text{ ft}} = 9.002 \times 10^{-4}$$

or

$$\left(\frac{e}{d}\right)^{-1} = \frac{1}{\epsilon} = 1111$$

From the friction factor chart, Figure 5.25, for $1/\epsilon = 1111$ and $Re = 61,588$, we obtain

$$\lambda = 0.023$$

The mechanical friction loss may be computed from Equation (5.156):

$$\hat{F} = \frac{\Delta p_{fr}}{\varrho} = \lambda \frac{\ell}{d} \frac{w^2}{2}$$

$$= (0.023)\left(\frac{73 \text{ m}}{0.0508 \text{ m}}\right)\frac{(1.33 \text{ m/s})^2}{2}$$

$$= 29.23 \frac{\text{J}}{\text{kg}} \text{ or } 9.8 \frac{\text{ft-lb}_f}{\text{lb}_m}$$

Example 5.19

A liquid is flowing through a 250-mm i.d. commercial steel pipe ($e = 2.5$ mm) with a head loss of 5.3 m over a length of 150 m. The kinematic viscosity of the process liquid is 2.3×10^{-6} m²/s. Determine the volumetric flow rate.

Solution

We may use Equation (5.156) $[h_{fr} = \lambda(\ell/d)(w^2/2g)]$ to solve for w:

$$e/d = \frac{2.5}{250} = 0.010$$

or

$$(e/d)^{-1} = \frac{1}{\epsilon} = 100$$

As λ is unknown, a trial and error solution for w must be made. An initial guess for λ is made. Assume $\lambda = 0.04$. Then,

$$5.3 \text{ m} = 0.04 \frac{150 \text{ m}}{0.25 \text{ m}} \frac{w^2}{2(9.806 \text{ m/s}^2)}$$

or

$$w = 2.08 \text{ m/s}$$

The Reynolds number computed from this velocity is

$$Re = \frac{wd}{\nu} = \frac{(2.08 \text{ m/s})(0.25 \text{ m})}{2.3 \times 10^{-6} \text{ m}^2/\text{s}} = 226{,}207$$

From the friction factor chart (Figure 5.25), this gives $\lambda = 0.037$. Interpolation between the assumed λ and the value obtained from Figure 5.25 gives a new $\lambda = 0.038$. Repeating the above calculations yields $w = 2.14$ m/s and a Reynolds number of 232,000. From Figure 5.25, Re of 232,000 reveals $\lambda \cong 0.038$.

Example 5.20

Oil is to be transferred at a rate of 28.9 m³/hr through a horizontal wrought iron pipeline. The head of fluid available to overcome the friction loss is 7.3 m for a 530-m length of pipe. The density and viscosity of the liquid are 890 kg/m³ and 3.8 cp, respectively. The roughness of the pipe to be used is 2.1×10^{-3} in. Determine the size of the pipe required to convey the oil.

Solution

The following information is given:
- $\varrho = 890$ kg/m³
- $\mu = 3.8$ cp $= 3.8 \times 10^{-3}$ kg/m-s
- $\ell = 530$ m
- $e = 2.1 \times 10^{-3}$ in $= 5.33 \times 10^{-5}$ m
- $Q = 28.9$ m³/hr $\times$ hr/3600 s $= 8.03 \times 10^{-3}$ m³/s

And the friction loss, $\hat{F} = gh_{fr} = (9.80665 \text{ m/s}^2)(7.3 \text{ m}) = 71.59$ J/kg. The pipe diameter d is unknown, which means we also do not know the area of the pipe or the oil's velocity:

$$F = \frac{\pi d^2}{4}$$

$$w = \left(8.03 \times 10^{-3} \frac{\text{m}^3}{\text{s}}\right)\left(\frac{4}{\pi d^2 \text{m}^2}\right) = \frac{0.01022}{d^2} \text{ m/s}$$

Again we may apply Equation (5.126):

$$h_{fr}g = \lambda \frac{\ell}{d}\frac{w^2}{2}$$

However, a trial and error solution is required because λ and Re are functions of d and w.

For an initial guess, let $d = 0.1$ m. Then,

$$\frac{1}{\epsilon} = \frac{d}{e} = \frac{0.1 \text{ m}}{5.33 \times 10^{-5} \text{ m}} = 1876$$

$$w = \frac{0.01022}{(0.1)^2} = 1.02 \text{ m/s}$$

$$Re = \frac{dw\varrho}{\mu} = \frac{(0.1)(1.02)(890)}{3.8 \times 10^{-3}} = 23{,}936$$

From Figure 5.25, for $Re = 23{,}936$ and $1/\epsilon = 1876$, we obtain a value for the friction factor, $\lambda = 0.026$. Hence,

$$h_{fr}g = (0.026)\left(\frac{530 \text{ m}}{0.1 \text{ m}}\right)\frac{(1.02 \text{ m/s})^2}{2} = 71.68 \text{ J/kg}$$

This agrees very well with the available head needed ($\hat{F} = 71.59$ J/kg). Hence, $d = 0.1$ m.

Table 5.3. *Computer Program for Solution to Example 5.20.*

```
10 DIM D(50)
20 REM PRGRM. TO CMPT. PIPE DIA
30 REM S=RHO/L=LENGTH/Q=FLOW/E=
   ROUGH/V=VISCOSIITY/F=FRICT./
   D=DIA./R=REYNOLDS/W=VELOCITY
   /A=AREA/G
40 READ S,L,Q,E,V,F,N
50 D(1)=.08
60 FOR I=1 TO N
70 A=.7854*D(I)^2
80 W=Q/A
90 E1=E/D(I)
100 R=D(I)*W*S/V
110 R1=E1/3.7+(6.81/R)^.9
120 G=- 5/LOG(R1)
130 F1=G*L*W^2/(2*D(I))
140 X=100*(F-F1)/F
150 IF X>-2 AND X<=2 THEN 180
160 D(I+1)=D(I)+.002
170 NEXT I
180 PRINT "AT ITERATION",I,"PIPE
     DIA. DETERMINED IN (M)=",D(
    I)
190 DATA 890,530..00803,.0000533
    ,.0038,71.6,25
200 END

AT ITERATION           22
PIPE DIA. DETERMINED IN (M)=
    122
```

Table 5.4. Computer Program Computations for Example 5.20.

I	Re	Computed Head Loss (J/kg)	Difference in Head Loss (%)
1	29,932.4	582.6	−713.67
2	29,202.3	515.7	−620.21
3	28,507.0	457.8	−539.39
4	27,844.1	407.6	−469.27
5	27,211.2	363.9	−408.21
6	26,606.5	325.7	−354.89
7	26,028.1	292.2	−308.17
8	25,474.4	262.9	−267.12
9	24,943.6	237.0	−230.94
10	24,434.6	214.1	−198.99
11	23,945.9	193.8	−170.68
12	23,476.4	175.8	−145.54
13	23,024.9	159.8	−123.17
14	22,590.5	145.5	−103.21
15	22,172.1	132.7	−85.36
16	21,769.0	121.3	−69.37
17	21,380.3	111.0	−55.02
18	21,005.2	101.7	−42.11
19	20,643.0	93.4	−30.47
20	20,293.1	85.9	−19.97
21	19,954.9	79.1	−10.46
22	19,627.8	72.9	−1.86

AT ITERATION 22

PIPE DIA. DETERMINED IN (M) = 0.122

If the computed head did not agree, a second guess for d would have been necessary and the computations would have to have been repeated.

Alternative Solution

An alternative solution to Example 5.20 would be to program the equations using a convergence scheme. Equation (5.162) could be used to compute λ instead of using Figure 5.25. Such a computer program in BASIC language is listed in Table 5.3. Table 5.4 lists the results of several iterations performed by the program until convergence on the head loss is achieved. Note that the computer program predicts a required pipe diameter of 0.12 m. The difference between the two solutions is due to the inaccuracy in reading Figure 5.25.

The above friction factor expressions are based on fluids flowing under isothermal conditions. When the fluid is either heated or cooled through the pipe wall, viscosity changes over the flow cross section because of temperature gradients. Thus, the velocity field is modified and λ varies, especially in the laminar region where the temperature gradients are higher than in turbulent bulk flow.

For all the friction factor expressions given, excluding Equation (5.164), correction factors must be applied. These corrections are based on the assumption that the fluid temperature can be represented by a mean bulk temperature (defined as the arithmetic average of the inlet and outlet temperature). The correction factors, denoted by ψ, are as follows:

$$\text{For } Re > 2100: \psi = \begin{cases} \left(\dfrac{\mu}{\mu_w}\right)^{0.17} & \text{for heating} \\ \left(\dfrac{\mu}{\mu_w}\right)^{0.11} & \text{for cooling} \end{cases}$$

$$\text{For } Re < 2100: \psi = \begin{cases} \left(\dfrac{\mu}{\mu_w}\right)^{0.38} & \text{for heating} \\ \left(\dfrac{\mu}{\mu_w}\right)^{0.23} & \text{for cooling} \end{cases}$$

where

μ = viscosity of fluid at the average bulk temperature
μ_w = viscosity at the temperature of the pipe wall

To obtain the friction factor corresponding to the fluid's average bulk temperature, divide the isothermal λ value by ψ.

Flow through Varying Cross Sections

If the flow suddenly undergoes a change in cross-sectional area, additional head losses result. Figure 5.26 illustrates one example in which the flow experiences a sudden expansion. The head loss generated from this case is due to shocks as the flow from the smaller cross section impacts against a slower-moving fluid in the larger conduit.

The force of the rapid flow pF_2 acts against the plane F_2 in the larger section

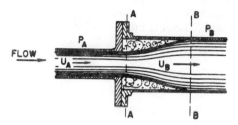

FIGURE 5.26. Flow through a sudden expansion.

and in the opposite direction to the force from the slower-moving flow p_1F_2. The momentum change $d\phi$ with time dt is the following:

$$d\psi = (p_1F_2 - p_2F_2)dt \tag{5.165}$$

The fluid momentum over time dt is the product of mass and velocity. Hence, the change in momentum is

$$(w_2 - w_1)dM = (w_2 - w_1)\frac{\gamma}{\varrho}(F_2w_2)dt \tag{5.166}$$

where F_2W_2 is the volumetric flow rate.

According to the second law of mechanics, Equations (5.165) and (5.166) may be equated:

$$\frac{p_1 - p_2}{\gamma} = \frac{w_2(w_2 - w_1)}{g} \tag{5.167}$$

This expression shows that the total pressure change is due to resistance and changes in kinetic energy.

From Equation (5.167), it is determined that $p_2 > p_1$. Thus, despite resistances, the pressure in the larger pipe in Figure 5.26 will be greater than in the smaller one, resulting in a reversal of fluid motion, i.e., swirling occurs in the expansion's "corners." From the total pressure change, Equation (5.167) may be used to calculate local resistances. For turbulent flow, the Bernoulli equation may be applied to the section preceding the expansion in the following form:

$$\frac{p_1}{\gamma} + \frac{w_1^2}{2g} = \frac{p_2}{\gamma} + \frac{w_2^2}{2g} + h_\ell \tag{5.168}$$

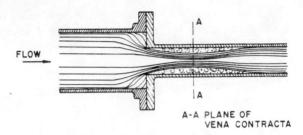

FIGURE 5.27. Flow through a sudden contraction.

Substituting for $p_1 - p_2$ from Equation (5.167) into the above expression, we solve for resistance h_ℓ in units of height of liquid column:

$$h_\ell = \frac{(w_1 - w_2)^2}{2g} \qquad (5.169)$$

Equation (5.169) is valid for any cross section of piping and is also applicable to liquid discharge from pipes into large vessels. For the latter case, $w_2 \ll w_1$, so resistance is equal to the kinetic energy of the flow in the piping, $h_\ell = w^2/2g$. Experimental observations for turbulent flows of both compressible and incompressible fluids have verified Equation (5.169).

The effects of sudden contractions on flows are discussed by Mott (1972), Cheremisinoff (1979, 1981a,b), and Simpson (1969). Figure 5.27 illustrates flow through a sudden contraction. As shown, the stream cannot flow around sharp corners. The flow's cross section first contracts and then expands to fill the cross section of smaller piping. Because of these phenomena, energy losses c· resistances arise, which may be estimated from the Weisbach equation

$$h_\ell = \left[0.04 + \left(\frac{1}{a} - 1\right)^2\right] \frac{w_2^2}{2g} = \psi \frac{w_2^2}{2g} \qquad (5.170)$$

where a is the *contraction loss coefficient*, i.e., the ratio of the minimum flow cross section to the cross section of the smaller pipe, and W_2 is the average velocity in the smaller section.

Table 5.5. *Coefficients of Contraction [for use in Equation (5.170)].*

F_1/F_2	0.01	0.1	0.2	0.4	0.6	0.8	1.0
a	0.6	0.61	0.62	0.65	0.7	0.77	1.0
ψ	0.5	0.46	0.42	0.33	0.23	0.13	0.0

The coefficients a and ψ depend on the ratio of the pipes' cross sections, F_2/F_1. Typical values are given in Table 5.5. Coefficient ψ also may be calculated from the following equation:

$$\psi = \frac{1.5\left(1 - \dfrac{F_2}{F_1}\right)}{3 - \dfrac{F_2}{F_1}} \qquad (5.171)$$

These equations may be used to calculate the resistances encountered in liquid discharge from large vessels through orifices or from vessels into piping. For the latter situation, F_2/F_1 in Equations (5.171) can be ignored. Recall that Equation (5.170) only expresses the head loss due to the change in flow cross section. The total head loss also accounts for the change in kinetic energy in accordance with the Bernoulli equation.

Example 5.21

Water is flowing at a rate of 1800 gal/hr through a distribution system. At one point in the system, a 1-in pipeline is suddenly expanded to 1.75 in i.d.

1 Determine the energy loss that occurs due to the sudden enlargement.
2 Determine the difference between the pressure immediately ahead of the sudden enlargement and the pressure downstream from the enlargement.

Solution

Part 1

$$d_1 = 1 \text{ in} = 0.0833 \text{ ft}; \quad F_1 = \frac{\pi}{4}(0.0833)^2 = 0.00545 \text{ ft}^2$$

$$d_2 = 1.75 \text{ in} = 0.1458 \text{ ft}; \quad F_2 = \frac{\pi}{4}(0.1458)^2 = 0.01670 \text{ ft}^2$$

Hence,

$$w_1 = \frac{Q}{F_1} = \frac{1800 \text{ gph}}{0.00545 \text{ ft}^2} \times \frac{1 \text{ ft}^3/\text{s}}{449 \text{ gpm}} \times \frac{\text{hr}}{60 \text{ min}} = 12.3 \text{ fps}$$

$$w_2 = \frac{Q}{F_2} = \frac{1800}{0.01670} \times \frac{1}{449} \times \frac{1}{60} = 4.0 \text{ fps}$$

Using Equation (5.169),

$$h_\ell = \frac{(w_1 - w_2)^2}{2g} = \frac{(12.3 - 4.0)^2 [\text{ft}^2/\text{s}^2]}{2(32.2 \text{ ft}/\text{s}^2)} = 1.07 \text{ ft}$$

This means that 1.07 ft-lb$_f$ of energy is dissipated from each lb$_m$ of water that flows through the sudden enlargement.

Part 2

To evaluate the pressure differential, we apply the Bernoulli equation in the following manner:

$$\frac{p_1}{\varrho} + z_1 + \frac{w_1^2}{2g} - h_\ell = \frac{p_2}{\varrho} + z_2 + \frac{w_2^2}{2g}$$

Solving for $p_1 - p_2$,

$$p_1 - p_2 = \varrho[(z_2 + z_1) + (w_2^2 - w_1^2)/2g + h_\ell]$$

If we assume the expansion is horizontal, then $z_1 - z_2 = 0$. Hence,

$$p_1 - p_2 = 62.4 \frac{\text{lb}}{\text{ft}^3}\left(0 + \frac{(4.0)^2 - (12.3)^2}{(2)(32.2)} \text{ ft} + 1.07 \text{ ft}\right)$$

$$= 62.4(0 - 2.10 + 1.07) \text{lb}/\text{ft}^2$$

$$= -64.3 \frac{\text{lb}}{\text{ft}^2} \times \frac{1 \text{ ft}^2}{144 \text{ in}^2} = -0.447 \text{ lb}_f/\text{in}^2$$

Hence, p_2 is greater than p_1 by 0.447 psi.

Example 5.22

By means of a momentum balance and mechanical energy balance, develop an expression for the loss that occurs for liquid flowing through the sudden expansion shown in Figure 5.26.

Solution

To write the momentum balance we select a control volume so as not to include the large pipe wall. The boundaries selected are defined by planes 0 and 2 in Figure 5.26. We may assume that the flow through plane 0 occurs only through an area F_1. The frictional drag force is neglected, and all the loss is assumed to be from the eddies within this volume. We note, therefore, that $p_0 = p_1$, $w_0 = w_1$, and $F_0 = F_2$. Making a momentum balance between planes 0 and 2,

$$p_0 F_2 - p_2 F_2 = M w_2 - M w_0$$

where the mass flow rates are

$$M = w_1 \varrho F_1$$

And from continuity:

$$w_2 = (F_1/F_2) w_1$$

Substituting these terms into the momentum balance,

$$(p_1 - p_2) F_2 = w_1 \varrho F_1 \left[\frac{F_1}{F_2} w_1 - w_1 \right]$$

Noting that

$$\frac{F_1}{F_2} = \frac{\frac{1}{4} \pi (d_1)^2}{\frac{1}{4} \pi (d_2)^2} = \left(\frac{d_1}{d_2} \right)^2$$

and rearranging, we obtain

$$\frac{p_2 - p_1}{\varrho} = w_1^2 \left(\frac{d_1}{d_2} \right)^2 \left[1 - \left(\frac{d_1}{d_2} \right)^2 \right]$$

Applying the mechanical energy balance equation over planes 0 and 2,

$$\frac{w_1^2 - w_2^2}{2} - \Sigma \hat{F} = \frac{p_2 - p_1}{\varrho}$$

Equating the last two expressions and solving for the friction loss, we obtain

$$\Sigma \hat{F} = \frac{w_1^2}{2} (1 - \beta)^2$$

where $\beta = (d_1/d_2)^2$.

The reader should develop a derivation in a similar manner for flow through a sudden contraction (see Figure 5.27).

Comparable losses occur when flows undergo gradual changes in cross section (Figure 5.28). For a smooth conical expansion, as shown in Figure 5.28(a), where the cone's apex angle $\beta < 10°$, head losses may be estimated from the Eligner equation:

$$h_\ell = \left(\frac{F_2}{F_1} - 1\right)^2 \sin\beta \left(\frac{w_1^2}{2g}\right) \tag{5.172}$$

For $\beta > 10°$, head losses may be estimated from the expression for a sudden expansion.

For a gradual expansion, as shown in Figure 5.28(b), where β is in the range of 7 to 35°, head losses may be computed from the following empirical expression:

$$h_\ell = 0.35 \left(\log \frac{\beta}{2}\right)^{1.22} \frac{(w_1 - w_2)^2}{2g} \tag{5.173}$$

For gradual expansions with $\beta < 7°$, head losses are determined by integration of the Bernoulli and Darcy-Weisbach equations.

At $\beta > 40°$, head losses may become very high and even exceed those encountered in sudden expansions. Note also that the flow may be turbulent in the small cross section but laminar in the large section.

For gradual contractions, as shown in Figure 5.28(c), head losses are insignificant (especially for a smooth contraction surface and large Reynolds numbers). Equation (5.170) may be used, setting $h_\ell = 0.05$, independent of the ratio F_2/F_1, provided that the flow is turbulent in the narrow cross section. If the flow in the contracted section is laminar (as in the case of discharge from a vessel), even when the resistance is small, an unusual pressure decrease occurs that does not correspond to Poiseuille's law. This decrease occurs over a length equivalent to 0.065 ReD (the entrance region of a pipe). The pressure gradient $p_0 - p_1$ for laminar flow at the entrance of a pipe of length L may be calculated from the data given in Table 5.6.

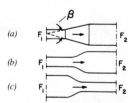

FIGURE 5.28. Flow in gradual expansions and contractions.

Table 5.6. Data for Estimating Pressure Losses.[a]

$\dfrac{L}{D}$ Re	0.005	0.01	0.02	0.03	0.04	0.05	0.06
$\dfrac{p_0 - p_1}{\gamma \dfrac{w^2}{2g}}$	2.1	2.6	3.4	4.1	4.7	5.3	6.0

[a] D = pipe diameter; L = pipe length; γ = liquid specific gravity [see Krylov (1947) for further details].

Example 5.23

Develop an expression for the average upstream velocity for liquid flow through a horizontal gradual contraction shown in Figure 5.28(c). Assume the liquid to be of constant density. The expression should be written in terms of the pressure differential $P_1 - P_2$ across the contraction.

Solution

Assign upstream conditions as $P_1(\text{N/m}^2)$, $F_1(\text{m}^2)$, and $w_1(\text{m/s})$ and downstream conditions as $P_2(\text{N/m}^2)$, $F_2(\text{m}^2)$, and $w_2(\text{m/s})$. As density is constant, $\varrho_1 = \varrho_2 = \varrho$. And, from the mass balance continuity equation,

$$w_2 = \frac{w_1 F_1}{F_2}$$

where $F_1/F_2 = (d_1/d_2)^2$.

As this is a horizontal flow system, $z_1 = z_2 = 0$, the Bernoulli equation may be written as follows:

$$\frac{w_1^2}{2} + \frac{P_1}{\varrho} = \frac{w_1^2 (F_1/F_2)^2}{2} + \frac{P_2}{\varrho}$$

or

$$\frac{\varrho w_1^2}{2}[(F_1/F_2)^2 - 1] = P_1 - P_2$$

Hence,

$$w_1 = \sqrt{\frac{P_1 - P_2}{\varrho} \frac{2}{(\beta^4 - 1)}}$$

in SI units, or

$$w_1 = \sqrt{\frac{P_1 - P_2}{\varrho} \frac{2g_c}{(\beta^4 - 1)}}$$

in English units.

Example 5.24

Estimate the energy loss that occurs when 20 gpm of water flows from a 1-in tube into a 2.9 in tube through a gradual enlargement having an included angle of 30°.

Solution

$$Q = 20 \text{ gpm} \times \frac{\text{ft}^3}{7.48 \text{ gal}} \times \frac{\text{min}}{60 \text{ s}} = 0.0446 \text{ ft}^3/\text{s}$$

$$F_1 = \frac{\pi}{4}\left(\frac{1}{12}\right)^2 = 0.00545 \text{ ft}^2, \quad w_1 = \frac{0.0446}{0.00545} = 8.2 \text{ fps}$$

$$F_2 = \frac{\pi}{4}\left(\frac{2.9}{12}\right)^2 = 0.0459 \text{ ft}^2, \quad w_2 = \frac{0.0446}{0.0459} = 0.97 \text{ fps}$$

Applying Equation (5.173),

$$h_\ell = 0.35 \left(\log \frac{\beta}{2}\right)^{1.22} \frac{(w_1 - w_2)^2}{2g}$$

$$= 0.35 \, (\log 15)^{1.22} \frac{(8.2 - 0.97)^2}{2 \times 32.2} = 0.36 \text{ ft (or 0.16 psi)}$$

Alternative Solution

Another approach to estimating friction losses for any type of expansion or contraction is to apply Equation (5.169) with an appropriate discharge coefficient. That is,

$$h_\ell = C_L \frac{(w_1 - w_2)^2}{2g}$$

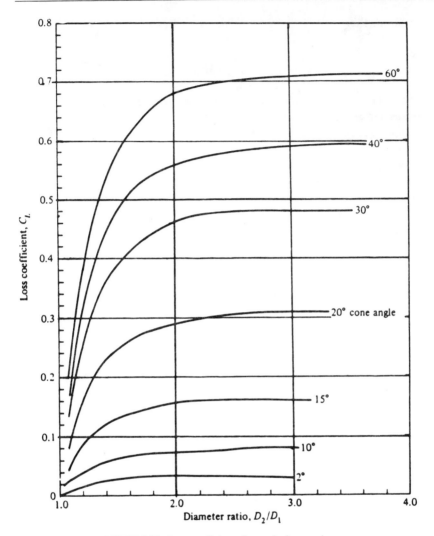

FIGURE 5.29. Loss coefficients for gradual expansions.

Discharge coefficients for various cross-sectional changes are reported in the literature [see Streeter and Wylie (1970) or Perry (1950)]. Figure 5.29 provides a plot of discharge coefficient versus diameter ratio for gradual enlargements.

Then, $d_2/d_1 = 2.9/1 = 2.9$ and, from Figure 5.29 (for $\beta = 30°$),

$$C_L \cong 0.47$$

$$h_\ell = 0.47 \frac{(8.2 - 0.97)^2}{2 \times 32.2} = 0.38 \text{ ft } (0.17 \text{ psi})$$

Note that often it is assumed for quick estimates that $w_1 \gg w_2$. Then,

$$h_\ell = C_L \frac{w_1^2}{2g} = 0.47 \frac{(8.2)^2}{2 \times 32.2} = 0.49 \text{ ft } (0.21 \text{ psi})$$

The latter estimate is actually preferred when determining energy requirements for pumps. This provides a safety factor when selecting and sizing pump requirements.

Example 5.25

A large water holding tank is drained from a side nozzle 1.5-in in diameter. Determine the energy loss as the water flow undergoes a sudden contraction. The water discharges at a rate of 27 gpm.

Solution

Energy losses may be estimated for the sudden contraction from the Weisbach equation [Equation (5.170)]:

$$h_\ell = \left[0.04 + \left(\frac{1}{a} - 1\right)^2\right]\frac{w_2^2}{2g} = \psi \frac{w_2^2}{2g}$$

where

$\psi = 1.5(1 - F_2/F_1)/(3 - F_2/F_1) \cong 0.5$ (as $F_1 \gg F_2$)
$Q = 27$ gpm $\times$ cfs/449 gpm $= 0.0602$ cfs
$F_2 = \pi/4(1.5/12)^2 = 0.0123$ ft²; $w_2 = 0.0602/0.0123 = 4.91$ fps

and

$$h_\ell = 0.5 \frac{(4.91)^2}{2 \times 32.2} = 0.187 \text{ ft (or } 0.081 \text{ psi)}$$

Flow through Piping Components

When fluids flow through fittings such as bends, elbows, etc., the flow direction is altered and the action of centrifugal force generates flow patterns that aggravate the bulk liquid motion. This phenomenon creates additional head losses that are best described for engineering calculations in terms of equivalent lengths of straight pipe with pipe diameters having the same head loss as the fittings. Beij (1938) first described such losses for fluid flow through

Table 5.7. Friction Loss Coefficients for Turbulent Flow through Values and Fittings.

Type Fitting or Valve	Head Loss Coefficient, K_e	L_e/D
Elbow, 45°	0.35	17
Elbow, 90°	0.75	35
Tee	1	50
Return Bend	1.5	75
Coupling	0.04	2
Union	0.04	2
Gate Valve		
Wide open	0.17	9
Half open	4.5	225
Globe Valve		
Wide open	6.0	300
Half open	9.5	475
Angle valve		
Wide open	2.0	100
Check Valve		
Ball	70.0	3500
Swing	2.0	100
Water Meter, disk	7.0	350

90° pipe bends. Head losses in fittings and valves may be expressed by a simple relationship of the following form:

$$h_\ell = K_e \frac{w_1^2}{2} \qquad (5.174)$$

where K_e is the head loss coefficient, which is a function of the ratio of the fittings' equivalent length to diameter ratio, L_e/D. Table 5.7 gives typical values of K_e for different valves and fittings. Note that w_1 is the velocity in the pipe immediately upstream of the fitting. Application of Equation (5.174) is limited exclusively to turbulent flows.

To evaluate total head loss for a piping system, equivalent lengths L_e of fittings and components are added to straight lengths of pipe in the system. That is, the total piping over which losses occur $= L_e +$ length of straight pipe. The Darcy equation then can be applied to obtain total head losses for a system:

$$\hat{F}_f = \frac{\Delta P_f}{\varrho} = 4\lambda_F \frac{\Sigma L_e}{D} \frac{w^2}{2} \qquad (5.175)$$

Table 5.8. Values for Parameter X in Equation (5.176)
[data of Konakov (1949)].

R/r	2	4	6	8	10	20	32
X	4	1.2	1.2	1.7	2.2	4.8	7.6

where λ_F is the Fanning friction factor, i.e., one-fourth the value obtained from Figure 5.25.

For turbulent flow through copper bends, Konakov (1949) recommends the following formula:

$$\frac{L_e}{D} = 0.0202 \, Xa^{1.10} Re^{0.032} \qquad (5.176)$$

where a is the bending angle in grades, and X is an empirical function of the ratio of bending radius R to pipe radius r. Values for X are given in Table 5.8.

Values given in Table 5.8 indicate that L_e/D is, for practical purposes, independent of the Reynolds number under turbulent conditions, and that the minimum resistance occurs at $R/r = 5$. However, for laminar flow, equivalent length L_e is a function of Re. For 90° elbows, the data of Wilson (1922) (Table 5.9) are recommended.

Example 5.26

Determine the equivalent length of pipe (in feet) of a wide-open angle valve positioned in a 6-in Schedule 40 pipe.

Solution

From Table 5.7, the equivalent length ratio L_e/D for a wide-open angle valve is 100. The inside diameter of a 6-in Schedule pipe is 0.5054 ft. Hence,

$$L_e = (L_e/D)(D) = 100 \times 0.5054 = 50.5 \text{ ft}$$

Example 5.27

Estimate the pressure loss across a half-open gate valve located in a horizontal 6-in Schedule 40 steel pipeline carrying 370 gpm of SAE 10W oil at 92°F.

Table 5.9. Equivalent Lengths of 90° Elbows for Laminar Flow
[data of Wilson (1922)].

Re	10	50	100	250	500	800	1000
L_e/D	2.5	3.5	6.0	12.0	22	27	30

Solution

To evaluate the pressure drop we shall apply the energy equation over a short section of piping containing the valve:

$$\frac{P_1}{\gamma} + z_1 + \frac{w_1^2}{2g} - h_\ell = \frac{P_2}{\gamma} + z_2 + \frac{w_2^2}{2g}$$

where subscripts 1 and 2 refer to points immediately before and after the valve, respectively. Note that h_ℓ is the minor loss, which, for this system, is the valve alone.

Solving the energy equation for the pressure drop, and as this is a horizontal system, $Z_1 = Z_2 = 0$:

$$P_1 - P_2 = \gamma \left(0 + \frac{w_2^2 - w_1^2}{2g} + h_\ell \right)$$

Now as the pipeline does not undergo a change in its cross section (except for the valve), we may assume that the flow quickly reaches steady state shortly after the valve and, hence, that $w_1 = w_2$. Therefore,

$$P_1 - P_2 = \varrho g h_\ell$$

The head loss can be determined from Equation (5.175):

$$h_\ell = 4\lambda_F \frac{\Sigma L_e}{D} \frac{W^2}{2g} = \lambda \left(\frac{\Sigma L_e}{D} \right) \frac{W^2}{2g}$$

From Table 5.7, for a half-open gate valve, $L_e/D = 225$. The actual diameter of a 6-in Schedule 40 pipe is 0.5054 ft. Hence,

$$F = \frac{\pi}{4} (0.5054)^2 = 0.201 \text{ ft}^2$$

and

$$w = \frac{370 \text{ gpm}}{0.201 \text{ ft}^2} \times \frac{1 \text{ ft}^3/\text{s}}{449 \text{ gpm}} = 4.1 \text{ fps}$$

The kinematic viscosity of the oil is $\nu = 1.7 \times 10^{-4}$ ft²/s. Hence, the Reynolds number is

$$Re = \frac{wD}{\nu} = \frac{(4.1)(0.5054)}{1.7 \times 10^{-4}} = 12{,}190$$

The roughness $e = 1.2 \times 10^{-4}$ ft. Therefore:

$$\epsilon = \frac{D}{e} = \frac{0.5054}{1.2 \times 10^{-4}} = 4212$$

From Figure 5.25, the friction factor λ is obtained; $\lambda = 0.031$ or $\lambda_F = 7.75 \times 10^{-3}$. Hence,

$$h_\ell = (0.031)(225)\frac{(4.1)^2}{(2)(32.2)} = 1.82 \text{ ft}$$

$$P_1 - P_2 = \gamma h_\ell = (0.875)(62.4)\frac{\text{lb}}{\text{ft}^3} \times 1.82 \text{ ft} \times \frac{\text{ft}^2}{144 \text{ in}^2} = 0.69 \text{ psi}$$

Thus, the pressure in the oil drops by 0.69 psi as it flows through the half-open valve. The student should repeat the problem for a fully open gate valve and a half-open globe valve. Compare results and comment.

Example 5.28

A large holding tank, 23 ft tall, is used to supply water to an irrigation ditch. The piping distribution for the system is shown in Figure 5.30. The farmer wants to maintain a flow rate of 150 gpm at the field but is not sure the tank is tall enough to maintain a water level to support that flow. Determine the level required in the tank to maintain a discharge rate of 150 gpm and whether the tank is tall enough. The pipe is made of cast iron with a roughness $e = 1.8 \times 10^{-4}$ ft.

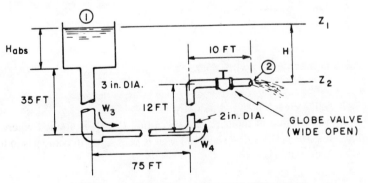

FIGURE 5.30. Water irrigation piping for Example 5.28.

Solution

The mechanical energy balance is written over points 1 and 2:

$$z_1 \frac{g}{g_c} + \frac{w_1^2}{2\alpha g_c} + \left(\frac{P_1}{\varrho_1} - \frac{P_2}{\varrho_2}\right) - \hat{W}_s = z_2 \frac{g}{g_c} + \frac{w_2^2}{2\alpha g_c} + \Sigma \hat{F}$$

where

$\varrho = 62.4$ lb/ft^3
$\mu = 0.35$ cp $= 2.3 \times 10^{-4}$ lb$_m$/ft-s
$Q = 150$ gpm $\times$ cfs/449 gpm $= 0.334$ ft^3/s

For the 3-in-diameter line,

$$F_3 = \frac{\pi}{4}\left(\frac{3}{12}\right)^2 = 0.0491 \text{ ft}^2; \quad w_3 = \frac{0.334}{0.0491} = 6.80 \text{ fps}$$

For the 2-in-diameter line,

$$F_4 = \frac{\pi}{4}\left(\frac{2}{12}\right)^2 = 0.0218 \text{ ft}^2; \quad w_4 = \frac{0.334}{0.0218} = 15.31 \text{ fps}$$

Total friction losses $\Sigma \hat{F}$ include:

1 Contraction loss at tank exit
2 Friction in the 3-in straight pipe
3 Friction loss in the 3-in elbow
4 Sudden contraction from 3-in to 2-in pipe
5 Friction in the 2-in straight pipe
6 Friction in 2-in elbows
7 Friction across the wide-open globe valve

We will evaluate each loss individually.
1. Contraction loss at the tank exit. Equation (5.170) is applied where it is assumed that the tank's cross-sectional area is very large in comparison to the 3-in-diameter exit. Hence,

$$\psi = 0.5$$

and

$$h_\ell = \psi \frac{w_3^2}{2g} = 0.5 \frac{(6.80)^2}{2 \times 32.2} = 0.359 \text{ ft or } 0.359 \text{ ft-lb}_f/\text{lb}_m$$

since $\Delta P/\varrho = h_\ell g/g_c$.

2. Friction in the 3-in pipe. First compute the Reynolds number:

$$Re = \frac{d_3 w_3 \varrho}{\mu} = \frac{\left(\frac{3}{12}\right)(6.80)(62.4)}{2.3 \times 10^{-4}} = 461{,}218$$

The flow is turbulent. Next, determine the relative roughness of the pipe:

$$\frac{d_3}{e} = \frac{3/12 \text{ ft}}{1.8 \times 10^{-4} \text{ ft}} = 1389$$

From Figure 5.25, the friction factor is determined to be $\lambda = 0.018$, or the Fanning friction factor is $\lambda_F = 4.5 \times 10^{-3}$. Applying Equation (5.175),

$$\hat{F}_f = 4\lambda_F \frac{\Sigma L_e}{D} \frac{w^2}{2} \cdot \frac{1}{g_c}$$

$$= 4(0.0045)\left(\frac{35 \text{ ft}}{0.25 \text{ ft}}\right) \frac{(6.80)^2[\text{ft}^2/\text{s}^2]}{2} \times \frac{\text{lb}_{f\text{-}s}^2}{32.174 \text{ ft-lb}_m}$$

$$\hat{F}_f = 1.81 \text{ ft-lb}_f/\text{lb}_m$$

3. Friction loss in the 3-in elbows. From Table 5.7 the head loss coefficient for a 90° elbow is $K_e = 0.75$. Applying Equation (5.174),

$$h_\ell = K_e \frac{w_3^2}{2g_c} = (0.75) \frac{(6.80)^2}{2(32.174)}$$

$$= 0.539 \frac{\text{ft-lb}_f}{\text{lb}_m}$$

4. Sudden contraction loss from 3-in to 2-in pipe. Applying Equations (5.170) and (5.171), where

$$F_4/F_3 = \frac{0.0218}{0.0491} = 0.444$$

Hence,

$$\psi = \frac{1.5(1 - 0.444)}{3 - 0.444} = 0.326$$

$$h_\ell = \psi \frac{w_4^2}{2g_c} = 0.326 \frac{(15.31)^2}{(2)(32.174)} = 1.187 \frac{\text{ft-lb}_f}{\text{lb}_m}$$

5. Friction in the 2-in straight pipe.

$$Re = \frac{d_4 w_4 \varrho}{\mu} = \frac{(2/12)(15.31)(62.4)}{2.3 \times 10^{-4}} = 692{,}300$$

and

$$\frac{d}{e} = \frac{0.167 \text{ ft}}{1.8 \times 10^{-4} \text{ ft}} = 928$$

From Figure 5.25, $\lambda = 0.020$ or $\lambda_F = 0.005$. Applying Equation (5.175), where the total length of 2-in piping is $\Sigma L = 75 + 12 + 10 = 97$ ft,

$$\hat{F}_f = 4\lambda_F \frac{\Sigma L}{D} \frac{w_4^2}{2g_c} = 4(0.005) \frac{(97)(15.31)^2}{2(32.174)} = 7.07 \frac{\text{ft-lb}_f}{\text{lb}_m}$$

6. Losses in the two 2-in elbows. From Table 5.7 $K_e = 0.75$. Applying Equation (5.174) again and noting there are two elbows,

$$h_\ell = 2K_e \frac{w_4^2}{2g_c} = (2)(0.75) \frac{(15.31)^2}{2(32.174)} = 5.46 \frac{\text{ft-lb}_f}{\text{lb}_m}$$

7. Friction loss across the globe valve. From Table 5.7, $L_e/D = 300$ for a fully open globe valve. As determined from item 5, $\lambda_F = 0.005$. Note that $w_2 = w_4$. Hence,

$$h_\ell = 4\lambda_F \frac{\Sigma L_e}{D} \frac{w_2^2}{2g_c} = 4(0.005)(300) \frac{(15.31)^2}{(2)(32.174)}$$

$$= 21.86 \frac{\text{ft-lb}_f}{\text{lb}_m}$$

Thus the total friction loss for the system is the sum of items 1–7:

$$\Sigma \hat{F}_f = 0.359 + 1.81 + 0.539 + 1.187 + 7.07 + 5.46 + 21.86$$

$$= 38.29 \frac{\text{ft-lb}_f}{\text{lb}_m}$$

Assigning the reference datum at point 2, then $z_2 = 0$ and $z_1 = H$. If the tank is large, we may assume $w_1 = 0$. Also, $w_2 = w_4 = 15.31$ fps. And, as the flow is turbulent, $\alpha = 1.0$.

Both the tank and the discharge pipe are open to the atmosphere, thus $P_1 = P_2$. Furthermore, it is reasonable to assume $\varrho_1 = \varrho_2$, so

$$\frac{P_1}{\varrho_1} - \frac{P_2}{\varrho_2} = \frac{P_1 - P_2}{\varrho} = 0$$

Finally, no work is performed on the fluid by a pump, so $\hat{W}_s = 0$. Substituting these values into the energy equation,

$$H \frac{g}{g_c} + 0 + 0 - 0 = 0 + \frac{(15.31)^2}{(2)(1)(32.174)} + 38.29$$

$$H = 41.93 \text{ ft}$$

Hence, the water level in the tank must be maintained at a height of 41.93 ft above the discharge outlet, or the absolute tank level should be

$$H_{abs} = H - (35 - 12) = 41.93 - (23) = 18.9 \text{ ft}$$

Therefore, the tank will not overflow.

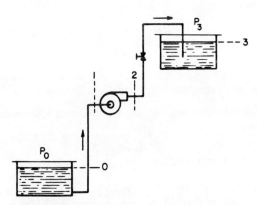

FIGURE 5.31. Pumping system for Example 5.29.

Example 5.29

Water at $t = 20°C$ is being pumped from an open tank located 5 m below the pump to another open tank positioned 20 m above the pump. The pipe is 52.5 mm in diameter and the system is shown in Figure 5.31. The horizontal pipe section in which the pump is positioned is 100 m long, and the horizontal section above the elevated tank is 10 m long. A fully open globe valve is on the discharge line. The efficiency of the pump is 75%. Determine the pump's power.

Solution

The volumetric rate is

$$V_{sec} = \frac{5000}{1000 \times 3600} = 0.00139 \text{ m}^3/\text{s}$$

Pipe cross-sectional area is

$$F = \frac{3.14 \times 0.0525^2}{4} = 0.00216 \text{ m}^2$$

The superficial water velocity in the piping is

$$w = \frac{V}{F} = \frac{0.00139}{0.00216} = 0.64 \text{ m/s}$$

The kinetic energy of 1 kg of water is

$$\frac{w^2}{2g} = \frac{0.64^2}{2 \times 9.81} = 0.021 \text{ m}$$

$t = 20°C$ and $\mu \cong 1$ cp, then $\mu g = 10^{-3}$ kg/m-s and the specific weight is $\gamma = 10^3$ kg/m³. Hence, the Reynolds number is

$$Re = \frac{wd\gamma}{\mu g} = \frac{0.64 \times 0.0525 \times 10^3}{10^{-3}} = 33,600$$

Since the flow is turbulent, we may use the following relationship to compute the friction factor:

$$\lambda = 0.0123 + \frac{0.7544}{Re^{0.38}}$$

$$\lambda = 0.0123 + \frac{0.7544}{33,600^{0.38}} = 0.0277$$

Table 5.10. Critical Reynolds Numbers for Coils.

D/d	15.5	18.7	50	2050
Re_{cr}	7600	7100	6000	2270

The actual length of piping from the bottom tank to the pump is 107 m. The piping has one 90° elbow, which has an L_e/D of 35 from Table 5.7. Consequently, the total equivalent piping length before the pump is

$$L = 107 + 35 \times 0.0525 = 108.8 \text{ m}$$

Applying the Darcy-Weisbach equation, the resistance before the pump is

$$Z_{0,1} = \lambda \frac{LW^2}{2gd} = 0.027 \frac{108.8}{0.0525} \times 0.021$$

$$= 1.2 \text{ m}$$

The length of the piping after the pump is 30 m. This section has two elbows and a valve having an $L_e/D = 300$ (from Table 5.7). Hence, the equivalent length after the pump is

$$L = 30 + 0.0525 (2 \times 30 + 300) = 49 \text{ m}$$

The resistance after the pump will be

$$Z_{2,3} = 0.0277 \times \frac{49}{0.0525} \times 0.021 = 0.54 \text{ m}$$

The pressure difference before and after the pump is

$$\frac{P_0 - P_1}{\gamma} = (Z_3 - Z_0) + Z = 27.0 + 1.2 + 0.54$$

$$= 28.7 \text{ m}$$

because $P_3 = P_0 = 1$ atm, $Z = 0$, and $Z_3 = 27$ m. Consequently,

$$P_2 - P_1 = 28.7 \times 10^3 \text{ kg/m}^2$$

Hence, the pump's power is

$$N = \frac{(P_2 - P_1)V}{\eta} = \frac{(28.7 \times 10^3)0.00139}{0.75}$$

$$= 53.2 \frac{\text{kg-m}}{\text{s}}$$

or

$$N = \frac{53.2}{102} = 0.53 \text{ kW (or 0.71 hp)}$$

Coils, often used in a variety of process equipment, have critical Reynolds numbers Re_{cr} very different from those encounterd in straight pipes. Here, the friction factor is a function of the ratio of coil diameter D to pipe diameter d. Typical critical values for different D/d ratios are given in Table 5.10.

Table 5.10 reveals that the smaller the coil diameter, the higher the critical value of the Reynolds number and, consequently, the longer the flow remains in the laminar regime. That is, the critical Re of 2100 marking the transition between laminar and turbulent flows through pipes is greatly exceeded for coils. The resistance of laminar flow in coils may be estimated from the Darcy-Weisbach equation,

$$h_\ell = - \frac{\Delta P}{\gamma} = \lambda \frac{L}{D} \frac{w^2}{2g}$$

using the following empirical friction factor expression developed by White (1929a,b):

$$\lambda = C \frac{64}{Re} \tag{5.177}$$

Coefficient C approaches unity with increasing coil radius of curvature. C may be computed from the following formula or from the values given in Table 5.11:

$$C = \frac{1}{1 - \left[1 - \left(\frac{11.6}{Re\sqrt{d/D}}\right)^{0.45}\right]^{1/0.45}} \tag{5.178}$$

For turbulent flow through coils of $D/d > 500$, resistances are comparable to

Table 5.11. Resistances in Coils [values computed from Equation (5.178)].

$Re\sqrt{d/D}$	10	50	100	250	400	600	1000	2000
C	1.0	1.2	1.5	2.0	2.5	3.0	4.5	5.0

those in straight pipes. At very high Reynolds numbers ($Re > 110,000$), Jeschke (1925) found that the resistance coefficient λ is independent of Re and could be expressed by the following:

$$\lambda = 0.0238 + 0.0891 \, d/D \qquad (5.179)$$

To estimate hydraulic resistances through coils, one may assume the coil to be a return bend and apply Equation (5.176) to obtain an L_e/D ratio.

Until now, the analyses presented have focused on evaluating flow resistances through piping of uniform or round cross sections. In practice, however, other configurations are encountered frequently. One example is an annular cross section in a heat exchanger, where the fluid must flow between two concentric pipes. The formulae and analyses presented are still valid for other configurations; however, the Reynolds number and, consequently, the friction factor, are redefined in terms of an equivalent diameter.

As previously noted, the *equivalent diameter* is defined as four times the hydraulic radius r_h, where r_h is the ratio of the cross-sectional area of flow and the wetted perimeter, i.e., the perimeter of the channel contacting the fluid. That is,

$$d_{eq} = 4r_h \qquad (5.180A)$$

and

$$r_h = \frac{F}{p} \qquad (5.180B)$$

where

F = cross-sectional area of channel
p = wetted perimeter of the channel

Table 5.12 gives specific formulas for the hydraulic radius for channels of various configurations.

For an elliptical cross section, coefficient k in Table 5.12 must be computed. k is a function of the ratio $\dot{s} = (a - b)/(a + b)$. Values are given in Table 5.13.

By redefining Re in terms of r_h, head losses for turbulent flows may be estimated from Equation (5.175) by virtually the same computation steps outlined in the example problems presented earlier. Unfortunately, this design method is not reliable for laminar flows.

There are two approaches to evaluating resistances for laminar flows

Table 5.12. Hydraulic Radii for Different Channel Configurations.

Cross Section	r_h
Circular pipe, diameter D	$D/4$
Annulus between two concentric pipes—D and d are the outside and inside diameters of the annulus	$\dfrac{D-d}{4}$
Rectangular duct with sides a and b	$\dfrac{ab}{2(a+b)}$
Square duct with a side a	$a/4$
Ellipse with axes a and b	$\dfrac{ab}{K(a+b)}$
Semicircle of diameter D	$D/4$
A shallow flat layer with depth h	h
Liquid film with thickness t on the vertical pipe of diameter D	$t - \dfrac{t^2}{D} \cong t$

through irregular configurations. In the first method, the Reynolds number is evaluated from an equivalent diameter defined by Equations (5.180A) and (5.180B), whence a friction factor may be determined from an expression of the following form:

$$\lambda = \frac{a}{Re} \qquad (5.181)$$

The value of a depends on the flow geometry. Othmer (1945) has given values for different configurations. These values are presented in Table 5.14, along with formulas for the equivalent diameter.

The second method for estimating losses in the laminar regime for non-circular cross sections is based on the derivation of Poiseuille-type equations. As an example, consider the flow configuration of an annular cross section.

Table 5.13. Coefficients for Evaluating the Hydraulic Radius of an Ellipse.

$\dot{s}$	0.2	0.3	0.4	0.5	0.6	0.7	0.8	0.9	1.0
k	1.010	1.023	1.040	1.064	1.092	1.127	1.168	1.216	1.273

Table 5.14. Equivalent Diameters and Values for Parameter a in Equation (5.181).

Cross Section	h/b	D_{eq}	a
Circular with diameter D	–	D	64
Square with side h	–	h	57
Triangle with side h	–	0.58h	53
Ring with width b	–	2b	96
Ellipse with axes h and b (h = small axis)	0.7	1.17h	65
	0.5	1.30h	68
	0.3	1.44h	73
	0.2	1.50h	76
	0.1	1.55h	78
Rectangle with sides h and b (h = small side)	1/∞	2h	96
	0.1	1.82h	85
	0.2	1.67h	76
	0.25	1.60h	73
	0.33	1.50h	69
	0.50	1.33h	62

The derivation of Poiseuille's law, presented earlier, results in an expression for the resistance per unit length of channel:

$$-\frac{dP}{dL} = \frac{32\mu\overline{W}}{D^2 + d^2 - \frac{D^2 - d^2}{\ln D/d}} \qquad (5.182)$$

where $\overline{W}$ is the average superficial velocity. For a rectangular section of sides a and b, Poiseuille's equation takes the following form:

$$-\frac{dP}{dL} = \frac{4\overline{W}\mu}{abn} \qquad (5.183)$$

n is a function of a/b; typical values are given in Table 5.15.

For an elliptical cross section, the head loss is expressed as

$$-\frac{dP}{dL} = \frac{4\overline{W}\mu(a^2 + b^2)}{a^2 b^2} \qquad (5.184)$$

where a and b are the semiaxes of an ellipse.

Table 5.15. Values of n in Equation (5.183) for a Rectangular Cross Section.

a/b	0.1	0.2	0.3	0.4	0.5	0.6	0.7	0.8	0.9	1.0
n	0.03	0.06	0.08	0.10	0.11	0.133	0.136	0.138	0.139	0.140

Finally, let us consider two infinitely wide plates with a distance of $2b$ separation. Poiseuille's equation may be written as follows:

$$-\frac{dP}{dL} = \frac{3\mu \overline{W}}{b^2} \quad (5.185)$$

The velocity distribution between the two parallel plates is given by the following equation:

$$w = \frac{3}{4} \frac{\Gamma(b^2 - y^2)}{b^3 \gamma} \quad (5.186)$$

where

Γ = weight rate per unit plate width
y = distance to the plane of symmetry (i.e., the plane in the middle between the plates)
γ = specific gravity

Velocity Dampening

For simple pipe flow, we have observed the formation of a characteristic velocity distribution across the area of flow, with a maximum occurring at the pipe axis and zero at the pipe wall. The velocity gradient, especially in laminar flow, may have a relatively high value. In some industrial applications it is desirable to have almost a flat velocity profile, for example, in gas cleaning applications and sedimentation.

The dampening or smoothing of the velocity profiles may be accomplished by one of two methods described by Stokes (1946). The first, shown in Figure 5.32(a), is based on a continuous flow constriction; the second [Figure

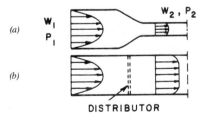

FIGURE 5.32. Two methods for dampening out velocity profiles.

5.32(b)] involves the use of a porous or a holed distributor. The advantage of the first method is that only a small additional resistance to the flow develops. Although the second method has a considerably higher head loss, costly fabrication and installation of the constriction are avoided. We shall analyze each approach from the principles already presented.

Applying the Bernoulli equation to the first method [Figure 5.32(a)], flow elements moving along the axis may be expressed as follows:

$$w_{max_2}^2 - w_{max_1}^2 = \frac{2g}{\gamma}(P_1 - P_2) \quad (5.187)$$

where subscripts 1 and 2 refer to conditions before and after the constriction, respectively. If mean velocity $\overline{W}_1$ is known, it is possible to obtain a similar relationship in terms of average velocities:

$$\overline{W}_2'^2 - \overline{W}_1^2 = \frac{2g}{\gamma}(P_1 - P_2) \quad (5.188)$$

$\overline{W}_2'$ is not equivalent to the velocity in the constricted piping, because the kinetic energy is determined from the average velocity $\overline{W}^2/2g\alpha$, which contains a coefficient α accounting for the change in local velocities. The average velocity in the reduced section, $\overline{W}_2$, will be somewhat different from $\overline{W}_2'$ in Equation (5.188).

Equating the Equations (5.187) and (5.188),

$$\overline{W}_2'^2(1 - \delta)^2 - \overline{W}_1^2 = w_{max_2}^2 - w_{max_1}^2 \quad (5.189A)$$

where

$$\hat{\delta} = \frac{\overline{W}_2 - \overline{W}_2'}{\overline{W}_2} \quad (5.189B)$$

From continuity, we note that

$$\overline{W}_1 F_1 = \overline{W}_2 F_2$$

where F_1 and F_2 are the cross-sectional areas of the large and constricted piping, respectively. Combining this with Equation (5.189A), the following relationship is obtained:

$$\frac{w_{max_2}}{\overline{W}_2} = \sqrt{1 + \left(\frac{F_2}{F_1}\right)^2\left[\left(\frac{w_{max_1}}{\overline{W}_1}\right)^2 - 1\right] - 2\hat{\delta} + \hat{\delta}^2} \quad (5.190)$$

Table 5.16. Dampening Rates Computed from Equation (5.191).

F_2/F_1	0	0.123	0.236	0.414	0.618	0.721	0.820	1.00
$\dfrac{w_{max_2}}{\overline{W}_1}$	1.0	1.023	1.080	1.23	1.46	1.60	1.74	2.00

Deviation of the ratio $w_{max_2}/\overline{W}_2$ from unity indicates the rate of velocity dampening in the constricted piping. Parameter δ in the expression is usually small and so may be neglected. Even for a very large velocity gradient the error incurred by neglecting δ is less than 5% (the worst case being $w_{max_1}/\overline{W}_1 = 2$). Hence, Equation (5.190) may be written as

$$\frac{w_{max_2}}{\overline{W}_2} = \sqrt{1 + \left(\frac{F_2}{F_1}\right)^2 \left[\left(\frac{w_{max_1}}{\overline{W}_1}\right)^2 - 1\right]} \qquad (5.191)$$

Equation (5.191) was applied to evaluating the most unfavorable velocity distribution, i.e., when $w_{max_1}/\overline{W}_1 = 2$. The rate of dampening computed from Equation (5.191) for different F_2/F_1 ratios is given in Table 5.16.

Values in Table 5.16 indicate that the velocity distribution may be assumed dampened for F_2/F_1 less than about 0.11. Note, however, that as the fluid moves away from the entrance of the constriction, the fluid nearest the wall slows down again and the normal velocity distribution is eventually reestablished.

The other method of velocity dampening involves inserting a perforated plate inside the piping [Figure 5.32(b)]. The average velocities before and after the plate are equal to each other. The pressure drop of a turbulent liquid flow through the disk is proportional to the fluid's kinetic energy:

$$\frac{P_1 - P_2}{\gamma} = K \frac{\overline{W}^2}{2g} \qquad (5.192)$$

The Bernoulli equation for liquid motion along the pipe axis is

$$\frac{P_1}{\gamma} + \frac{w_{max_1}^2}{2g} = \frac{P_2}{\gamma} + \frac{w_{max_2}^2}{2g} + K \frac{w_{max_2}^2}{2g} \qquad (5.193A)$$

Evaluating $P_1 - P_2$ and substituting into Equation (5.192), we obtain

$$\frac{w_{max_2}}{\overline{W}} = \sqrt{\frac{(w_{max_1}/\overline{W})^2 + K}{1 + K}} \qquad (5.193B)$$

This expression should be considered only as an approximation because the resistance coefficient K in Equation (5.192) is larger than the K value in (5.193A) and (5.193B). In most cases, however, the difference between the two leads to small errors. At any rate, Equation (5.193A) shows that at higher flow resistance (higher K) the velocity distribution is dampened more effectively, i.e., the ratio $w_{max_2}/\overline{W}$ is very close to unity.

FLOW NORMAL TO TUBE BANKS

Flow normal to a bank of tubes is encountered in a variety of industrial equipment, such as tubular heat exchangers and condensers. The magnitude of the flow resistance in such equipment is due to the contraction and expansion of the flow. This resistance also depends on the tube layout (e.g., staggered or unstaggered) relative to the flow direction, as illustrated in Figure 5.33. The critical Reynolds number for these flows has been determined as follows:

$$Re_{cr} = \frac{t' w_{max} \gamma}{\mu g} \cong 40 \tag{5.194}$$

where t' is the distance between tubes, and w_{max} is the maximum fluid velocity (in the narrow section).

For turbulent flow, i.e., $Re > 40$, flow resistance may be estimated from an expression of the following form:

$$h_r = \lambda N' \frac{w_{max}^2}{2g} \tag{5.195}$$

where

$h_r = \Delta P/\gamma$ = resistance of the liquid column
N' = number of tube rows in the direction of flow
λ = resistance coefficient depending on Re and the tube layout

The coefficient λ may be determined from the formulas reported by Jakob (1938).

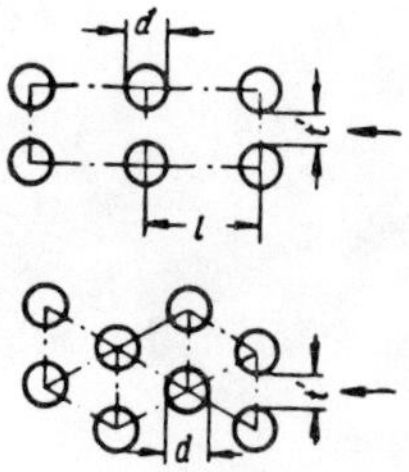

FIGURE 5.33. Flow normal to a bank of tubes. Flow resistance is a function of the tube layout: (a) an unstaggered layout; (b) a staggered layout.

Table 5.17. Resistance Coefficients for Banks of Tubes.

t'/d	0.2	0.5	1.0	1.5	2.0	5.0	10.0
λ	0.64	0.50	0.40	0.36	0.30	0.18	0.10

For a single row of tube layout, λ is determined from the following:

$$\lambda = \left[0.175 + \frac{0.32 \frac{\ell}{d}}{(t'/d)^n} \right] Re_d \tag{5.196}$$

where e, d, t' are defined in Figure 5.33, and $Re_d = t'w_{max}\gamma/\mu g = dw_{max}\gamma/\mu g$. At $Re_d \leq 10{,}000$ (still turbulent), λ is constant. Note that when $e = 2d$, $t' = d$, Equation (5.196) simplifies to

$$\lambda = 1.82\, Re^{-0.15} \tag{5.197}$$

where $Re = Re_d$. For rows of staggered tubes in the Reynolds number range of 2000 to 4000, the following equation is recommended:

$$\lambda = \left[0.92 + \frac{0.44}{(t'/d)^{1.08}} \right] Re_d^{-0.15} \tag{5.198}$$

Or, for the case of $t' = d$,

$$\lambda = 1.40\, Re^{-0.15} \tag{5.199}$$

Equations (5.196)–(5.199) give good estimates when the number of rows, N', exceeds 10. At $N' < 10$, actual head losses are somewhat higher than those predicted by applying these equations. At $N' = 4$, the error from applying these correlations is 7%; at $N' = 3$, 15%, and at $N' = 2$, up to 30%.

For flow around a single row of tubes ($N' = 1$), the data of Boucher (1947) should be used. Table 5.17 gives values of λ for different t'/d ratios at $Re = 10{,}000$ (recall that λ is constant at $Re \leq 10{,}000$ for a specific t'/d).

For laminar flow around tubes ($Re < 40$), the resistance may be calculated from an expression analogous to the Darcy equation [see Chilton (1938) for details]:

$$h_r = \lambda \frac{\ell}{d_{eq}} N' \frac{w_{max}^2}{2g} \tag{5.200}$$

where

$h_r = \Delta P/\gamma$ is the resistance of the liquid column
ℓ = distance between rows in the direction of flow
w_{max} = maximum velocity in the intertubular space
d_{eq} = equivalent diameter equal to four times hydraulic radius

Consider flow between two rows of tubes related to unity length of a tube, $\ell(t' + d) - \pi d^2/4$, where $\pi d^2/4$ is part of the volume occupied by corresponding segments of tubes. The surface of this tube is equal to πd; therefore, the equivalent diameter is

$$d_{eq} = 4 \frac{\ell(t' + d) - \dfrac{\pi d^2}{4}}{\pi d} = \frac{4\ell(t' + d)}{\pi d} - d \qquad (5.201)$$

We then define an equivalent Reynolds number:

$$Re_{eq} = \frac{d_{eq} w_{max} \gamma}{\mu g} \qquad (5.202)$$

For a row of staggered tubes, $Re_{eq} = 1 - 100$:

$$\lambda = \frac{106}{Re_{eq}} \qquad (5.203)$$

Substituting this expression for λ into Equation (5.200), we obtain

$$h_r = \frac{53\mu N' \ell w_{max}}{\gamma d_{eq}^2} \qquad (5.204)$$

For a row of unstaggered tubes, Bergelin (1949) found that λ values for turbulent flow multiplied by 1.5 gave good estimates. Applications involving flow around tubes are usually accompanied by heat transfer. The tube surface has a temperature different from the average or bulk temperature of the fluid. The effect of wall temperature is considered in the method of Siedez (1936), in which λ_{is}, corresponding to the temperature of the flow, is estimated first, and the true coefficient λ is determined from the following equation:

$$\lambda = \frac{\lambda_{is}}{a\left(\dfrac{\mu}{\mu_w}\right)^n} \qquad (5.205)$$

where μ is the viscosity at the temperature of the flow and μ_w is the viscosity at the temperature of the wall surface [see also Chilton (1938) for details].

For laminar flow, $a = 1.1$ and $n = 0.25$; for turbulent flow, $a = 1.0$ and $n = 0.14$.

OPTIMUM PIPE DIAMETER

A very practical consideration in formulating plant layouts is the determination of optimum pipeline sizes for specified flow conditions and pump energy limitations. The optimum is defined as the pipe size that provides an acceptable resistance per unit length within specified economic constraints. In other words, it is the most economical pipe diameter for a given set of operating conditions. As illustrated in Example 5.9, the problem is somewhat complicated by the fact that the Darcy equation [Equation (5.156)] cannot be applied in a straightforward manner. For a given volumetric flow, both the linear velocity w and Reynolds number (and, consequently, the friction factor) are functions of the pipe diameter.

Hence, the Darcy equation is an implicit function of the diameter, and direct determination of D, especially for turbulent flow, is often impossible. To address this problem, we begin by rewriting Darcy's equation in the following form:

$$-\frac{\Delta P}{\gamma} = \lambda \frac{8V^2 L}{D^5 \pi^2 g} \qquad (5.206)$$

Also, the Reynolds number is denoted as

$$Re = \frac{4V\gamma}{\pi \mu g D} \qquad (5.207)$$

Rearranging this last expression in terms of D and substituting into Equation (5.206), we obtain

$$\frac{128\ V^3 \gamma^4 \Delta P}{\pi^3 \mu^5 g^4 L} = \lambda Re^5 \qquad (5.208)$$

From a standard friction factor–Reynolds number chart such as in Figures 5.34 or 5.25, we may construct a plot of λRe^5 versus Re, where λRe is equivalent ot the LHS of Equation (5.208). From such a plot, a value for the Reynolds number corresponding to λRe^5 can be obtained. Once the Reynolds number is known, the pipe diameter may be computed [use Equation (5.207)].

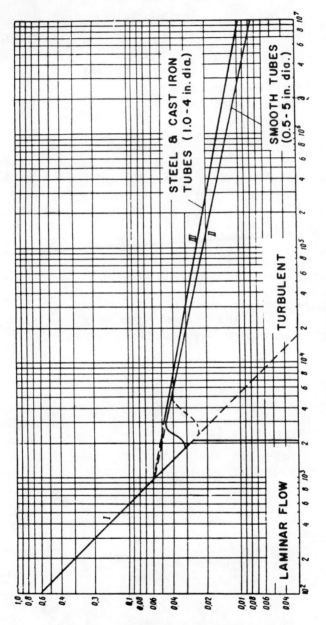

FIGURE 5.34. Expanded friction factor chart for steel and cast iron tubes and smooth tubes.

As is often the case, several pipe sizes will be considered acceptable in terms of head losses for a given flow rate; however, if we select a larger diameter, then construction costs will be high and, over a planned operating lifetime, maintenance, repair, and amortization expenses k_i will also be high. At the same time, a large diameter means lower hydraulic resistance and, consequently, lower power consumption for transportation. Hence, production expenses k_p will be relatively low.

Conversely, if we select a smaller diameter, hydraulic resistance increases along with production costs, whereas amortization expenses decrease. The relationships of k_i (maintenance/amortization costs) and k_p (production costs) to pipe diameter are shown in Figure 5.35. The sum of k_i and k_p equals the total yearly expenses for a pipe network. This also is shown in Figure 5.35 as a plot of Σk versus diameter. As shown, the relationship has a minimum that corresponds to the optimum pipe diameter.

Determination of the optimum pipe diameter may be formulated into a rigorous design procedure, which we shall outline and illustrate below. Both series of expenses should be related to a unit length of piping and a basis of one year's operation. The cost per unit length of piping is directly proportional to the diameter and may be expressed by the following relationship:

$$C_1 = XD^n \tag{5.209}$$

From data supplied by manufacturers and contractors on the costs of piping and installation, coefficients X and n can be evaluated.

Pipe components such as fittings and valves are also included in the analysis. The cost of fittings depends on the corresponding pipe diameter and may be considered a fractional cost of the piping, i.e., $C_2 = jC$, where j is a coefficient denoting fractional cost. The total cost per unit length C of a piping network, including fittings and valves, is

$$C = (1 + j)XD^n \tag{5.210}$$

where $C = C_1 + C_2$.

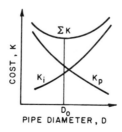

FIGURE 5.35. Relationship of operating costs and capital investment to pipe size.

By assigning an operating life for the system, e.g., 10–15 years, we may evaluate the portions of the total cost C, attributed to amortization expenses (denote as a) and to maintenance costs (b), such as repairs, painting, insulation, etc. That is, the total yearly expenses for amortization and maintenance are

$$k_i = (a + b)(1 + j)XD^n \qquad (5.211)$$

Production or operating expenses depend on the number of hours of operation per year (Y), the hourly rate of energy needed for fluid transportation ($\tilde{N}$), and the cost of a unit of energy (C_e):

$$k_p = Y\tilde{N}C_e \qquad (5.212)$$

The power required for fluid transportation may be expressed as $V\Delta P/\eta$, where V = volumetric rate, ΔP = pressure drop per unit length of pipe and η = the mechanical pump efficiency. Hence,

$$k_p = \frac{W_g \Delta p k C_e}{Y\eta} \qquad (5.213)$$

where W_g is the weight rate.

From Equation (5.206) with $L = 1$, we obtain

$$\Delta p = \frac{8}{\pi^2} \frac{\lambda W^2}{g\gamma D^5} \qquad (5.214)$$

For a Reynolds number range of 4000 to 2×10^7, the following equation may be used to obtain a friction factor:

$$\lambda = \frac{0.16}{Re^{0.16}} \qquad (5.215)$$

where

$$Re = \frac{GD}{\mu g} \qquad (5.216)$$

where

$$G = \frac{W}{\pi D^2/4}$$

Hence, the friction factor expression is

$$\lambda = \frac{0.16}{\left(\dfrac{W}{\pi D^2/4} \times \dfrac{D}{\mu g}\right)^{0.16}} = 0.154\left(\frac{D\mu g}{W}\right)^{0.16} \quad (5.217)$$

Substituting λ from Equation (5.217) into Equation (5.214), then into Equation (5.213), we obtain, after some simplification,

$$k_p = 0.125\,\frac{W^{2.84}\mu^{0.16}YC_e}{D^{4.84}\gamma^2\eta g^{0.84}} \quad (5.218)$$

Thus, Equations (5.218) and (5.211) provide estimates of production costs and amortization and maintenance expenses, respectively. The sum of these two is the total yearly cost for a piping system:

$$\Sigma k = (a + b)(j + 1)XD^n + 0.125\,\frac{W^{2.84}\mu^{0.16}YC_e}{D^{4.84}\gamma^2\eta g^{0.84}} \quad (5.219)$$

The optimum diameter is the minimum cost incurred for the pipe system. This minimum cost can be obtained by differentiating Equation (5.219) with respect to D and setting the derivative equal to zero:

$$\frac{d(\Sigma k)}{d(D)} = n(a + b)(j + 1)XD^{n-1} - \frac{0.605W^{2.84}\mu^{0.16}YC_e}{D^{5.84}\gamma^2\eta g^{0.84}} = 0$$

Thus, the optimum piping diameter is

$$D_0^{5.84} = \frac{0.605W^{2.84}\mu^{0.16}YC_e}{n(a + b)(j + 1)X\gamma^2\eta g^{0.84}} \quad (5.220\text{A})$$

For pipe diameters greater than ¾ in, exponent n in Equation (5.209) is, for practical purposes, unity. Hence, the final expression for the optimum pipe diameter is

$$D_0 = 0.918\left[\frac{YC_e}{(a + b)(j + 1)X\eta}\right]^{0.17}\frac{\mu^{0.027}}{g^{0.14}}\frac{W^{0.48}}{\gamma^{0.34}} \quad (5.220\text{B})$$

Examining this expression reveals that the dependency of D_0 on viscosity is very small ($\mu^{0.027}$). For example, in the range of 0.02 cp (the viscosity of air at STP) to 30 cp (a typical oil), $\mu^{0.027}$ changes by only a few percentage points. Hence, $\mu^{0.027}$ can be assumed as constant. And denoting

$$k_1 = 0.918 \left[\frac{YC_e}{(a+b)(j+1)X\eta}\right]^{0.17} \frac{\mu^{0.027}}{g^{0.14}} \qquad (5.221)$$

as a constant (actually assuming k_1 to be constant is a good approximation to within 10%), the optimum pipe diameter expression may be written simply as

$$D_0 = k_1 \frac{W^{0.48}}{\gamma^{0.34}} \qquad (5.222)$$

The derivation of Equation (5.222) is based on the condition of turbulent flow. Hence, the D_0 calculation should be checked by computing the Reynolds number and comparing against the criterion for turbulent pipe flow.

A similar design procedure is given by Normand (1948) for laminar flows. Following Normand, the derivation is the same through Equation (5.214). Departure from the above procedure comes from the use of the laminar friction factor $\lambda = 64/Re$, or from definition:

$$\lambda = 16\pi D\mu g/W$$

Repeating the same derivation procedure, but using the laminar λ expression, total expenses are

$$\Sigma k = (a+b)(j+1)XD^n + \frac{128}{\pi} \frac{W^2 YC_e\mu}{\gamma^2 D^4 \eta}$$

Again, differentiating and setting the derivative equal to zero, we obtain

$$D_0^{5+n} = \frac{512}{\pi} \frac{W^2 YC_e \mu}{\gamma^2 \eta n(a+b)(j+1)X} \qquad (5.223)$$

For $D_0 > \tfrac{3}{4}$ in, the exponent $n = 1$ and the expression simplifies to

$$D_0 = 2.77 \left[\frac{YC_e}{(a+b)(j+1)X\eta}\right]^{0.2} \left(\frac{W^2\mu}{\gamma^2}\right)^{0.2} \qquad (5.224)$$

Denoting

$$k_2 = 2.77 \left[\frac{YC_e}{(a+b)(j+1)X\eta}\right] \qquad (5.225)$$

as a constant, we obtain the following expression for calculating the optimum pipe diameter for laminar flow:

$$D_0 = k_2 \left(\frac{W^2 \mu}{\gamma^2}\right)^{0.2} \quad (5.226)$$

For laminar flow, $Re = GD_0/\mu g < 2100$. Or, substituting $W/(\pi D^2/4)$ for G,

$$Re = WD_0 \Big/ \frac{\pi D_0^2}{4} \mu g < 2100 \quad (5.227)$$

Replacing D_0 in this expression with Equation (5.226), we obtain

$$W_g < 2.2 \times 10^5 (k_2 g)^{1.67} \frac{\mu^2}{\gamma^{0.67}} = W_{g_{cr}} \quad (5.228)$$

This last expression evaluates the flow regime that will exist for the optimum pipe diameter. Hence, from information on the weight rate W_g, viscosity μ, specific weight γ, and computed constant k_2 [from Equation (5.225)], the proper expression for calculating the optimum pipe diameter—Equation (5.222) or (5.226)—can be selected. The following example problem illustrates the design procedure.

Example 5.30

Determine the optimum pipe size for transporting 5000 kg/hr of water. The minimum life of the pipeline may be assumed to be 10 years. Yearly maintenance and repair expenses are expected to be 5% of the initial piping costs. The cost of fittings and valves is roughly 10% of the cost of the pipe. The system is to be designed for 24-hour operation throughout the entire year. The pump efficiency η is 60%, and the cost of electrical power is \$0.2/kWh. Table 5.18 provides quotations on costs per unit length of piping for different pipe diameters.

Solution

We will apply Equation (5.228) to evaluate the flow regime for the most economical or optimum pipe size and, depending on the results obtained, use either Equation (5.226) or (5.222) to compute D_0.

Table 5.18. Cost per Unit Length ($/m) of Different Pipe Diameters for Example 5.30.

D (in)	D (mm)	Price ($/m)
¾	21.5	3.12
1	27.0	4.30
1¼	35.75	4.94
1½	41.25	5.80
2	52.5	7.51
2½	68	9.04
3	80.25	11.30
3½	92.5	13.7
4	105	15.4
5	130	19.9
6	155.5	23.7

The cost per unit length of piping is described by Equation (5.209): $C_1 = XD^n$, where $n \cong 1$. Plotting the data given in Table 5.18 and evaluating the slope (Figure 5.36), we find $X \cong 150$ $/m². Hence,

$$C_1 = 150\ D(\$/m)$$

Electrical costs are

$$C_e = 0.02\ \$/\text{kWh} = \frac{0.2}{3.67 \times 10^5} = 5.45 \times 10^{-7}\ \$/\text{kg}$$

Operating time is

$$Y = 365 \times 24 \times 3600 = 3.15 \times 10^7\ \text{s/yr}$$

The amortization expenses are $a = 0.10$.

Maintenance and repair costs are 5% of the pipe costs (or $b = 0.05$). Hence,

$$a + b = 0.10 + 0.05 = 0.15$$

The cost of all fittings is 10% of the capital costs for piping. Hence, $j = 0.1$. From these values, the term common to Equations (5.221), (5.224), and (5.228) may be evaluated:

$$\frac{YC_e}{(a+b)(j+1)X\eta} = \frac{(3.15 \times 10^7)5.45 \times 10^{-7}}{0.15 \times 1.1 \times 150 \times 0.6} = 1.20$$

Using Equation (5.228), we evaluate the transition from laminar to turbulent flow:

$$\mu = \frac{1}{9.8 \times 10^3} = 1.02 \times 10^{-4}\ \frac{\text{kg·s}}{\text{m}^2}$$

$$k_2 = 2.77(1.20)^{0.2} = 2.9$$

Hence, the critical weight velocity for flow through the optimum pipe size is

$$W_{g,cr} = 2.2 \times 10^5 (2.9 \times 9.81)^{1.67} \frac{(1.02 \times 10^{-4})}{1000^{0.67}}$$

$$= 7.35 \times 10^3 \text{ kg/s}$$

For this problem, $W_g = 5000/3600 = 1.39$ kg/s. As $W_{g,cr} > W_g$, the flow is turbulent and Equation (5.222) should be used to compute D_0.

From Equation (5.221),

$$k_1 = 0.918(1.20)^{0.17} \frac{(1.02 \times 10^{-4})^{0.027}}{9.81^{0.14}} = 0.545$$

Finally, from Equation (5.222),

$$D_0 = 0.545 \frac{1.39^{0.48}}{(1000)^{0.34}} = 6.1 \times 10^{-2} \text{ m}$$

FLOW OF LIQUID FILMS

Vertical Film Flow

The flow of relatively thin liquid films in either a cocurrent or countercurrent mode with gases or vapors is utilized in a variety of unit operations, such as absorption, distillation, humidification and a variety of cooling techniques. To a large extent, the performance of equipment utilizing film flow depends on the flow regime, the film thickness, and the fluid velocity.

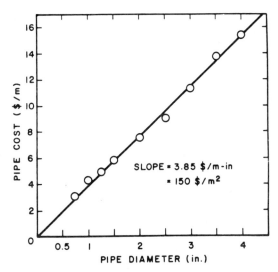

FIGURE 5.36. Plot of pipe costs versus diameter (prepared from data in Table 5.18).

Depending on the "film Reynolds number," three fundamental types of flow regimes are observed:

1. laminar liquid film flow with a smooth interface ($Re_f < 30$ to 50)
2. laminar flow with a rippled interface (30 to 50 $< Re_f <$ 100 to 400)
3. turbulent flow ($Re_f >$ 100 to 400)

The film Reynolds number is defined as

$$Re_f = \frac{w d_{eq} \varrho}{\mu} \quad (5.229)$$

where

w = film's average velocity
d_{eq} = equivalent diameter of the film

Re_f may also be expressed as follows:

$$Re_f = \frac{4\Gamma}{\mu} \quad (5.230)$$

where Γ is the mass rate of flow per unit width of wall.

This last definition allows the evaluation of the film Reynolds number without the need for a separate measurement of the film thickness and average velocity. The Equation (5.230) was derived based on the following reasoning. The equivalent diameter d_{eq} is defined as

$$d_{eq} = 4 \times \frac{\text{cross-sectional area of a film}}{\text{wetted perimeter of channel}} = \frac{4S}{p} = \frac{4p\delta}{p} = 4\delta \quad (5.231)$$

where δ = film thickness

Substituting for d_{eq} from Equation (5.231) into Equation (5.229) yields

$$Re_f = \frac{4w\delta\varrho}{\mu} \quad (5.232)$$

Denoting

$$\Gamma = w\delta\varrho$$

we obtain the modified Equation (5.230) for the film Reynolds number.

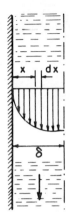

FIGURE 5.37. Liquid film flowing down a vertical wall. The downward arrows denote the velocity distribution.

Let us now consider laminar flow of a liquid film of constant thickness down a vertical wall. If the film is sufficiently thin, the flow may be assumed planar, even if it is flowing over a curved surface, e.g., flowing down the walls of a tube. Hence, the liquid interface is strictly parallel to the surface of the wall.

Figure 5.37 shows a portion of a liquid film having thickness δ and unit width flowing in the downward direction. For a Newtonian liquid, the shear stress at distance x from the wall is

$$\tau = -\mu \frac{dw}{dx} \tag{5.233}$$

where w is now the local linear velocity.

Similarly, the shear stress at distance $\chi + dx$ from the wall is

$$\tau + d\tau = -\mu \left[\frac{dw}{dx} + d\left(\frac{dw}{dx}\right) \right] \tag{5.234}$$

The motion of the film of thickness dx is caused by the resultant of these two stresses:

$$d\tau = -\mu d\left(\frac{dw}{dx}\right) \tag{5.235}$$

This downward-directed stress corresponds to the ratio of the resultant shear forces acting on the film interface and is equal to the weight of the film, i.e., the product of specific weight δ and the liquid volume $Ldx\cdot 1$, where L is the height of the wall. Hence, the resultant shear stress is

$$d\tau = \frac{Ldx\gamma}{L} = \gamma dx \qquad (5.236)$$

Substituting for $d\tau$ from Equation (5.235) into the above expression, we obtain the following differential equation:

$$\frac{\gamma}{\mu} dx = -\left(\frac{dw}{dx}\right)$$

Integrating this expression once gives

$$\frac{\gamma}{\mu} x = -\frac{dw}{dx} + c_1 \qquad (5.237A)$$

and integrating a second time gives

$$w = -\frac{x}{\mu} \cdot \frac{x^2}{2} + c_1 x + c_2 \qquad (5.237B)$$

The constants of integration may be evaluated by making use of the following boundary conditions corresponding to the solid wall and the liquid–gas interface:

$$\left. \begin{array}{l} at\ x = 0,\ w = 0,\ c_2 = 0 \\ at\ x = \delta,\ dw/dx = 0 \end{array} \right\} \qquad (5.237C)$$

Application of the boundary conditions yields

$$c_2 = 0$$

and

$$c_1 = \frac{w_{max}}{\delta} + \frac{\gamma}{\mu} \frac{\delta}{2}$$

Hence, an expression for the velocity distribution is

$$w = \frac{w_{max} x}{\delta} + \frac{\gamma}{2\mu} (\delta x - x^2) \qquad (5.238)$$

This expression is not readily used because it contains the maximum velocity w_{max}. Therefore, we differentiate this expression and obtain

$$\frac{dw}{dx} = \frac{w_{max}}{\delta} + \frac{\gamma}{2\mu}(\delta - 2x) \qquad (5.239)$$

Clearly, the velocity gradient at the interface is zero, i.e., at $x = \delta$, $dw/dx = 0$, because of the absence of shear stresses. Hence, Equation (5.239) reduces to an expression for the maximum velocity:

$$w_{max} = \frac{\delta^2 \gamma}{2\mu} \qquad (5.240)$$

Substituting this expression for w_{max} into Equation (5.238) gives the velocity distribution for a thin liquid layer flowing down a vertical wall due to the influence of gravity force alone.

$$w = \frac{\gamma}{\mu}\left(\delta x - \frac{x^2}{2}\right) \qquad (5.241)$$

This is our familiar parabolic expression, the profile of which is shown in Figure 5.37. The volumetric rate per unit width of film is obtained as follows:

$$V = \int_0^{\delta} w\,dx = \frac{\gamma}{\mu}\frac{\delta^3}{3} \qquad (5.242)$$

The average film velocity is obtained by dividing V by the film thickness:

$$\overline{W} = \frac{\gamma \delta^2}{3\mu} \qquad (5.243)$$

Comparison of Equations (5.243) and (5.240) reveals that **the maximum velocity on the surface of the layer is 1.5 times higher than the average velocity.** Following Bird et al. (1960), the film thickness δ may be expressed in terms of the average velocity, the volumetric flow rate, or the mass rate per unit width of the wall, $\Gamma = \varrho \delta w$; hence,

$$\delta = \sqrt[3]{\frac{3\mu \overline{W}}{\varrho g}} = \sqrt[3]{\frac{3\mu Q}{\varrho g}} = \sqrt[3]{\frac{2\mu \Gamma}{\varrho^2 g}} \qquad (5.244)$$

Equation (5.243) accurately predicts the *film thickness* for laminar flow of low-viscosity liquids (<1 cp) up to a critical Reynolds number ($4\Gamma/\mu$) in the range of 1000 to 2000.

Higher-viscosity liquids (10–20 cp) have film thicknesses appreciably below the predictions of Equation (5.244). This is especially true at and beyond the inception of surface wave motion, as observed by Jackson (1955). The inception of wave action has been observed to appear when the Froude number ($V^2/g\delta$) exceeds unity. It follows from this that the Reynolds number at which waves start depends only on the kinematic viscosity of the vertical flowing liquid. This assumes that the surrounding gas is stagnant and of negligible density. Then $\Gamma/\varrho' = 3(\mu'/\varrho')$, which corresponds to a wave inception Reynolds number of $Re_f = 12$.

Early studies by Friedman and Miller (1941) showed the ratio w_{max}/w_{avg} to be 1.5 up to $Re_f = 25$, after which it increased to about 2.2 at $Re_f = 100$. Jackson found w_{max}/w_{avg} to be 1.5 up to $Fr_f = 1$ ($Re_f = 12$), increasing rapidly to 2.2 at $Fr_f = 9.0$ ($Re_f = 108$) and slowly decreasing to about 1.8 at $Fr_f = 200$ ($Re_f = 2400$).

Example 5.31

A liquid having a kinematic viscosity of 3×10^{-4} m²-s⁻¹ and a density of 0.73×10^3 kg-m⁻³ is under gravity flow down a vertical wall. Determine the mass rate of flow for a measured film thickness of 3.2 mm.

Solution

An expression for the film thickness of a laminar flowing film under the influence of gravity is given by Equation (5.244). Rearranging this expression to solve for the mass flow rate per unit width of wall,

$$\Gamma = \frac{\delta^3 \varrho g}{3\nu} = \frac{(3.2 \times 10^{-3})^3 (0.73 \times 10^3)(9.80)}{3(3 \times 10^{-4})} = 0.261 \text{ kg-m}^{-1}\text{-s}^{-1}$$

To ascertain whether the flow is laminar, we compute the Reynolds number based on the calculated mass flow rate:

$$Re = \frac{4\Gamma}{\mu} = \frac{4\Gamma}{\varrho \nu} = \frac{4(0.261)}{(0.73 \times 10^3)(3 \times 10^{-4})} = 4.8$$

As the Reynolds number is below the observed upper limit for laminar flow, the calculated value of Γ is correct.

Example 5.32

An experiment was devised to check the accuracy of the laminar film flow expressions presented above. The liquid film thickness was obtained experimentally by weighing an amount of water flowing down a wall 2 m high by 1.2 m wide. The measurements showed the average amount of liquid on the wall to be 395 g for a mass flow of 52 kg/hr. Compare the measured value to the predicted.

Solution

$$\Gamma = \frac{G}{L} = \frac{52}{3600 \times 1.2} = 0.012 \text{ kg/m-s}$$

The average film velocity is

$$\overline{W} = \frac{\Gamma}{\rho\delta}$$

and the Reynolds number is (the viscosity of water is 1 cp or 10^{-3} N-s/m² at 20°C)

$$Re_f = \frac{\overline{W}\delta}{\nu} = \frac{\Gamma}{\mu} = \frac{0.012}{10^{-3}} = 12$$

To evaluate the flow regime, compare the obtained Reynolds number ($Re_f = 12$) with the flow's critical value:

$$Re_{cr} = 2.4\left(\frac{\sigma^3}{g\rho^3\nu^4}\right)^{1/11} = 2.4\left[\frac{(7.26 \times 10^{-2})^3}{9.81 \times 1000^3(10^{-6})^4}\right]^{1/11} \cong 22$$

As $Re < Re_{cr}$, the flow is laminar and the film thickness δ may be estimated from Equation (5.244):

$$\delta = \sqrt{\frac{3\mu\Gamma}{\rho^2 g}} = \sqrt{\frac{1.2 \times 10^{-2} \times 3 \times 10^{-3}}{9.81 \times 1000^2}} = 1.54 \times 10^{-4} \text{ m}$$

The amount of water retained on the wall when $\delta = 1.54 \times 10^{-4}$ m is

$$Q' = S\delta\rho = 1.2 \times 2 \times 1.54 \times 10^{-4} = 0.369 \text{ kg}$$

Comparing the experimental value of 0.395 kg with the prediction shows a relative error of about 7%.

Table 5.19. Physical Properties Data for 60% H_2SO_4 Solution.

t°, C	μ(cp)	ϱ(g/cm³)
0	10	1.52
10	8.0	1.51
20	6.5	1.50
30	5.5	1.49
40	4.5	1.48
50	3.7	1.47
60	3.2	1.47
70	2.7	1.46
80	2.3	1.45
90	1.9	1.44
100	1.6	1.43

Example 5.33

A system for contacting sulfuric acid with a process gas stream was proposed in the form of a vertical film flow system. A prototype was built to study the hydrodynamics. For this purpose, water was used. For a water weight rate of 3.3 kg/s-m, the film thickness δ was measured to be 1.0 mm. Using dimensional analysis, determine under what conditions a 60% solution of sulfuric acid will flow with a film thickness of 1.5 mm. Table 5.19 provides data on viscosities (cp) and densities (g/cm³) for the 60% H_2SO_4 solution at different temperatures. It may be assumed that the gas stream moves cocurrent with the downward liquid flow and at a much slower rate.

Solution

Film thickness δ is a function of the weight rate per unit width of a wall Γ, viscosity μ, specific gravity γ, and acceleration due to gravity g:

$$\delta = f(\Gamma, \mu, \gamma, g)$$

Rewriting this function in series,

$$\delta = a\Gamma^a \mu^b \gamma^c g^d + a'\Gamma^{a'} \mu^{b'} \gamma^{c'} g^{d'} + \cdots$$

or

$$\Sigma a = \left(\frac{\Gamma^a \mu^b \gamma^c g^d}{\delta} \right) = 1$$

The units on each of the variables are

$$\Gamma \equiv \frac{kg}{m\text{-}s}$$

$$\gamma \equiv \frac{kg}{m^3}$$

$$\mu \equiv \frac{kg\text{-}s}{m^2}$$

$$g \equiv \frac{m}{s^2}$$

$$\delta \equiv m$$

The expression inside the brackets must be dimensionless. Therefore, we may write the following identity:

$$\left(\frac{kg}{m\text{-}s}\right)^a \left(\frac{kg\text{-}s}{m^2}\right)^b \left(\frac{kg}{m^3}\right)^c \left(\frac{m}{s^2}\right)^d m^{-1} = kg^0 m^0 s^0$$

or

$$kg^{a+b+c} m^{-a-2b-3c+d-1} s^{-a+b-2d} = kg^0 m^0 s^0$$

Comparing the corresponding exponents we obtain three equations:

$$a + b + c = 0$$

$$-a + b - 2d = 0$$

$$-a - 2b - 3c + d - 1 = 0$$

Thus, there are three equations with four unknowns. Solving in terms of one of the unknowns, we obtain

$$a = a$$

$$b = \frac{2}{3} - a$$

$$c = -\frac{2}{3}$$

$$d = \frac{1}{3} - a$$

The obtained results are substituted into the above series expression:

$$\Sigma a \Gamma^a \mu^{2/3-a} \gamma^{-2/3} g^{1/3-a} \delta^{-1} = 1$$

This equation is then transformed by exponents:

$$\Sigma a \left(\frac{\Gamma}{\mu g}\right)^a \left(\frac{\mu^2 g}{\gamma^2 \delta^3}\right)^{1/3} = 1$$

We further change the expression by the general function describing the given process:

$$\frac{\Gamma}{\mu g} = \Phi\left(\frac{\mu^2 g}{\gamma^2 \delta^3}\right)$$

Instead of $\mu g/\gamma$, we introduce the kinematic viscosity coefficient ν, then

$$\frac{\Gamma}{\mu g} = \Phi\left(\frac{\delta^3 g}{\nu^2}\right)$$

Assume

$$\delta = 0.001 \text{ m}$$

$$g = 9.81 \text{ m/s}^2$$

$$\mu = 1 \text{ cp (for water)}$$

Then,

$$\mu g = 10^{-3} \text{ kg/m-s}$$

and,

$$\nu = \frac{\mu g}{\gamma} = \frac{10^{-3}}{10^3} = 10^{-6} \text{ m}^2/\text{s}$$

Consequently,

$$\frac{\delta^3 g}{\nu^2} = \frac{(10^{-3})^3 \times 9.81}{(10^{-6})^2} = 9.81 \times 10^3$$

For the H_2SO_4 solution, this fraction will have the same numerical value, provided the proper kinematic viscosity is known. According to the problem statement, the film thickness should be 1.5 mm. That is, the kinematic viscosity coefficient should be

$$\nu = \sqrt{\frac{\delta^3 g}{9.81 \times 10^3}} = \sqrt{\frac{(1.5 \times 10^{-3})^3 \times 9.81}{9.81 \times 10^3}} = 1.84 \times 10^{-6} \text{ m}^2/\text{s}$$

From Table 5.19, values of the coefficient of kinematic viscosity are computed:

T,°C: 0   10   20   30   40   50   60   70   80   90   100

$\nu \times 10^{-6}$ m²/s):   6.6   5.3   4.35   3.7   3.05   2.52   2.18   1.85   1.59   1.32   1.12

We see that at $T = 70°C$ and $\nu = 1.84 \times 10^{-6}$ m²/s the dimensionless fraction on the right-hand side of the expression $\delta^3 g/\nu^2$ will be the same as that for H_2O. Only at this temperature is it possible for the acid solution to have the same velocity as the flowing water. That is,

$$\left(\frac{\Gamma}{\mu g}\right)_{H_2O} = \left(\frac{\Gamma}{\mu g}\right)_{H_2SO_4}$$

For water, $\Gamma = 3.3$ kg/m-s, and $\mu g = 10^{-3}$ kg/m-s:

$$\frac{\Gamma}{\mu g} = \frac{3.3}{10^{-3}} = 3300$$

For H_2SO_4 at temperature $T = 70°C$, the viscosity is $\mu = 2.7$ cp; therefore, $\mu g = 2.7 \times 10^{-3}$ kg/m-s and

$$\Gamma = 3300 \, \mu g = 3300 \times 2.7 \times 10^{-3} = 8.9 \text{ kg/m-s}$$

Hence, the acid must be fed to the system at a rate of 8.9 kg/m-s and at a temperature of 70°C.

Wavy Flow

In the above analyses it was implied that the gas stream flowed cocurrent to the falling liquid film and at a relatively low velocity. This made a description of the hydrodynamics of the film strictly in terms of the liquid motion possible. Even in the case of countercurrent flow (liquid flowing vertically downward and gas upward), the film thickness will be independent of the gas stream velocity, provided the gas rate is low. If, however, the gas velocity is increased, the friction or shear force at the film interface, acting in the opposite direction to the liquid flow, becomes significant, and the film motion slows down.

At a certain gas velocity ($\sim$ 5–10 m/s), an equilibrium state arises between gravitational and friction forces. This condition leads to flooding, i.e., an accumulation of liquid in the column, eventually initiating entrainment and a dramatic increase in the hydraulic resistance. At still higher gas velocities, the film's flow direction becomes inverted and begins to "creep up," that is, the liquid and gas move upward cocurrently. When this occurs, the hydraulic resistance is first observed to decrease comparatively to the flooding regime but then increases. At gas velocities greater than 15 m/s, the liquid is entrained. In contrast, at very high gas velocities in the downward cocurrent flow case, the gas greatly increases the liquid film's velocity while decreasing its thickness.

With this background, we will now extend the analysis of constant-thickness laminar film flow down a vertical wall to include interaction with the gas phase. The equation of motion for this system takes the form

$$\gamma' - \gamma'' + \mu' \frac{d^2 w'}{dy^2} = 0 \qquad (5.245A)$$

where γ', γ'' are the specific gravities of liquid and gas, respectively, and μ' is the liquid viscosity.

Boundary conditions for this differential equation are

$$y = 0, \; w' = 0$$
$$y = \delta, \; \mu' \frac{dw'}{dy} = \pm \zeta'' \frac{\gamma'' w_2''^2}{2g} \qquad (5.245B)$$

where ($''$) refers to the gas phase and ζ is the resistance coefficient.

After integration and evaluation with the boundary conditions, we obtain

$$w' = \left(\pm \zeta \frac{\gamma'' w_r^2}{2g\mu'} + \frac{\gamma' - \gamma''}{\mu'} \right) y - \frac{\gamma' - \gamma''}{2\mu'} y^2 \qquad (5.246)$$

We shall assign the downward direction of flow as being positive. Hence, a plus sign is used on the second term in Equation (5.246) for the gas phase, i.e., the gas carries the film along in the same direction as the gravity force.

To indicate upflow of the gas (the gas retards the downflow of the film) a minus sign is used. The velocity of the film on the phase boundary is

$$w'_b = \pm \zeta'' \frac{\gamma'' w_r''^2}{2g\mu'} \delta + \frac{\gamma' - \gamma''}{2\mu'} \delta^2 \qquad (5.247)$$

where w_r'' is the relative velocity of the gas and is equal to $w'' - w'_b$.

The average (flow) velocity of the liquid film is

$$\overline{W}' = \frac{1}{d} \int_0^\delta w' \, dy = \pm \zeta'' \frac{\gamma'' w_r''^2}{4g\mu'} \delta + \frac{\gamma' - \gamma''}{3\mu'} \delta^2 \qquad (5.248)$$

In upward gas flow, the film is carried along upward and the liquid velocity becomes negative at

$$\frac{3\zeta'' w_r''^2 \gamma''}{4g\delta(\gamma' - \gamma'')} > 1 \qquad (5.249)$$

This inequality represents the criterion for "flooding."

When 30 to 50 $< Re_f <$ 100, a wavy regime is observed. That is, the interface is rippled and capillary waves appear at the interface that are occasioned by forces of gravity and viscosity acting on the film. The condition is illustrated in Figure 5.38.

Following the analysis of Kutateladze and Styrikovich (1958), we consider the limiting case that occurs when the gas velocity is close to zero. Hence, the

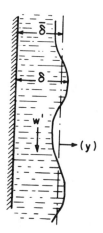

FIGURE 5.38. The formation of interfacial disturbances (waves) in film flow.

tangential stresses on the free surface may be neglected. From Equations (5.246) and (5.248), the velocity profile is

$$w'_x = \frac{3\overline{W}_x}{\delta}\left(y - \frac{y^2}{2\delta}\right) \tag{5.250}$$

During the wave motion, the average velocity along the cross section is a function of the x-coordinate and time t:

$$w'_x = \frac{3w'_x(x;t)}{\delta}\left(y - \frac{y^2}{2\delta}\right) \tag{5.251}$$

Substituting this expression for w'_x into Equation (5.250) and integrating over the limits of $y = 0$ and $y = \delta$, we obtain the following:

$$\frac{\partial \overline{W}'_x}{\partial t} + \frac{9}{10}\overline{W}'_x \frac{\partial \overline{W}'_x}{\partial x} = \frac{6}{\varrho'}\frac{d^3\delta}{dx^3} - \frac{3\nu \overline{W}'}{\delta^2} + \frac{\gamma' - \gamma''}{\varrho'} \tag{5.252}$$

For waves of small amplitude, the film thickness may be expressed as a binomial function:

$$\delta = \bar{\delta} + \psi'\bar{\delta} \tag{5.253}$$

where

$\bar{\delta}$ = the average thickness of the film
ψ' = the deviation coefficient of the instantaneous value of the thickness from its average value

Capillary waves originating on the surface of a film are not dampened. Therefore, all the terms of Equation (5.252) are functions of the x-coordinate and the phase velocity v, i.e., they are functions of the argument $(x - vt)$. Then,

$$\left.\begin{array}{l}\dfrac{\partial \delta}{\partial t} = -\bar{\delta}v\dfrac{\partial \varphi'}{\partial x} \\[1em] \dfrac{\partial \overline{W}'_x}{\partial t} = -v\dfrac{\partial \overline{W}'_x}{\partial x}\end{array}\right\} \tag{5.254}$$

Substituting these expressions into Equations (5.252) and (5.253), we have

$$\frac{9}{10}(\overline{W}'_x - v)\frac{\partial \overline{W}'_x}{\partial x} = \frac{6\bar{\delta}}{\varrho'} \times \frac{d^3\varphi'}{dx^3} - \frac{3v'\overline{W}'_x}{\bar{\delta}^2(1 + \varphi')^2} + \frac{\gamma' - \gamma''}{\varrho'} \qquad (5.255)$$

$$\frac{\partial}{\partial x}[(v - \overline{W}'_x)\bar{\delta}(1 + \varphi')] = 0 \qquad (5.256)$$

As follows from Equation (5.256),

$$\bar{\delta}(v - \overline{W}'_x)(1 + \varphi') = \text{const.} = \bar{\delta}(v - w'_0) \qquad (5.257)$$

where w'_0 is the average velocity in a cross section $\bar{\delta}$.

From this expression, the following relationship applies to waves at the film surface:

$$\overline{W}'_x = v - \frac{v - w'_0}{1 + \varphi'} \qquad (5.258)$$

Expanding as a series,

$$\left.\begin{array}{l}\overline{W}'_x = w'_0 + (v - w'_0)(\varphi' - \varphi'^2 + \varphi'^3 \ldots) \\[6pt] \dfrac{\partial \overline{W}'_x}{\partial x} = (v + w'_0)(1 - 2\varphi' + 3\varphi'^2 \ldots)\dfrac{d\varphi'}{dx}\end{array}\right\} \qquad (5.259)$$

Introducing Equation (5.259) into Equation (5.255) results in an expression for the dimensionless amplitude φ. Small amplitudes ($\varphi' \ll 1$) allow us to consider only first approximations for w'_x and $\delta \overline{W}'_x/\delta x$.

The film's energy loss due to friction per unit area of wall surface is

$$-\frac{dE}{dt} \approx \mu' \int_0^\delta \left(\frac{\partial w'_x}{\partial y}\right)^2 dy = 3\mu' \frac{\overline{W}'^2_x}{\delta} \qquad (5.260)$$

And, after averaging along the wavelength,

$$\left(\frac{d\overline{E}}{dt}\right)_{\lambda_0} = -\frac{3\mu'}{\lambda_0}\int_0^{\lambda_0} \frac{\overline{W}'^2_x}{\delta} dx \qquad (5.261)$$

where λ_0 denotes the wavelength at any given instance.

The average work per unit area of wall due to gravity is

$$\bar{L} = \Gamma$$

In a steady-state process, the condition $|dE/dt| = \bar{L}$ must be satisfied, whence it follows, after transformation, that

$$\bar{\delta}^3 = \frac{3\nu'\Gamma}{g(\gamma' - \gamma'')}\Phi \qquad (5.262)$$

where

$$\Phi = \frac{1}{\lambda_l}\int_0^{\lambda_l}\frac{\left(1 + \frac{\nu}{w_0'}\right)\varphi}{(1+\varphi)^3}dx \qquad (5.263)$$

Levich (1962) outlines the following relations as a second approximation:

$$\varphi = 0.21\sin[k(x - \nu t)]$$

$$\begin{cases} k = \left(0.9\dfrac{g\tilde{\Gamma}}{\gamma'\mu'}\right)^{1/2} \\ \nu = 2.4 w_0' \\ \Phi = 0.8 \end{cases} \qquad (5.264)$$

Investigations on wave profiles by Kapitsa and Zhetf (1948) showed good agreement between experiment and above theory.

Turbulent Film Flow

This last subsection examines the system of turbulent film flow of constant thickness on a vertical wall. The analysis closely follows that of Kutateladze and Styrikovich (1958). As a first approximation, turbulent flow may be divided into a laminar sublayer and a turbulent core. The generalized velocity profile may then be described by a system of two equations:

$$\left.\begin{array}{l}\tau_f \cong \mu'\dfrac{dw'}{dy} \; ; \text{ for } y < y_1 \\[2mm] \tau_f \cong \varrho' x^2 y^2 \left(\dfrac{dw'}{dy}\right)^2 \; ; \text{ for } y > y_1\end{array}\right\} \qquad (5.265)$$

where y_1 is the thickness of the laminar sublayer, and parameter χ is a constant that characterizes the structure of turbulence. Kutateladze and Styrikovich determined χ and y_1 to have values 0.4 and 11.6, respectively. Note that Equations (5.265), particularly the expression for the tangential stress due to turbulent friction, are only approximations.

From a steady flow balance, the tangential stresses in the liquid film may be expressed in terms of gravity force and the friction of the gas against the film interface:

$$\tau_f' = g(\varrho' - \varrho'')(\delta - y) \pm \zeta'' \frac{\varrho' w_r'^2}{2} \qquad (5.266)$$

At the wall, $y = 0$, and the tangential stresses become

$$\tau_w = g(\varrho' - \varrho'')\delta \pm \zeta'' \frac{\varrho' w_r'^2}{2} \qquad (5.267)$$

At the interface, $y = \delta$, hence,

$$|\tau_i| = \zeta'' \frac{\varrho' w_r'^2}{2} \qquad (5.268)$$

Because the laminar sublayer is relatively thin, tangential stresses in this region may be considered practically constant and equal to τ. If the gas moves slowly, i.e., $w_r'' \cong 0$, then we have the following expressions:

For the laminar sublayer,

$$g \frac{\gamma' - \gamma''}{\gamma'} \delta = \nu' \frac{dw'}{dy} \qquad (5.269)$$

For the turbulent core,

$$g\left(\frac{\gamma' - \gamma''}{\gamma'}\right)(\delta - y) = \chi^2 y^2 \left(\frac{dw'}{dy}\right)^2 \qquad (5.270)$$

After integrating Equations (5.269) and (5.270) and applying appropriate boundary conditions, the following expression is obtained for the velocity profile across the entire film:

$$w' = w_1' + \frac{1}{\chi}\sqrt{g\left(1 - \frac{\gamma''}{\gamma'}\right)}\left[2\sqrt{\delta - y} - 2\sqrt{\delta - y_1}\right.$$
$$\left. + \sqrt{\delta} \ln \frac{(\sqrt{\delta} - \sqrt{\delta - y})(\sqrt{\delta} + \sqrt{\delta - y_1})}{(\sqrt{\delta} + \sqrt{\delta - y})(\sqrt{\delta} - \sqrt{\delta - y_1})}\right]; \text{ for } y > y_1 \qquad (5.271)$$

In the vicinity of the solid wall, where $y \leq \delta$, it can be assumed that

$$\sqrt{1 - y/\delta} = 1 - \frac{y}{2\delta}$$

Hence, an approximate form of this expression is

$$w' \approx w_1' + \frac{1}{\chi}\sqrt{g\left(1 - \frac{\gamma''}{\gamma'}\right)}\left[\frac{y_1 - y}{\delta} + \ln\frac{y\left(2 - \frac{y_1}{2\delta}\right)}{y_1\left(2 - \frac{y}{2\delta}\right)}\right] \quad (5.272)$$

$$\cong w_1' + \frac{1}{\chi}\sqrt{g\left(1 - \frac{\gamma''}{\gamma'}\right)}\,\delta \ln y/y_1$$

This expression is the well-known logarithmic law of velocity distribution for turbulent flow in the vicinity of a solid wall. Note that in the derivation of this expression, the boundary conditions on the free surface of the film were ignored. Nevertheless, Equation (5.272) does provide good predictions of velocities in various cross sections of the film.

The average velocity in the film is thus obtained from

$$\overline{W}' = \frac{1}{\delta}\left(\int_0^{y_1} w'\,dy + \int_{y_1}^{\delta} w'\,dy\right) \quad (5.273A)$$

And, substituting the value w' from Equation (5.272) and assuming that $y_1 \ll \delta$, we obtain

$$\overline{W}' \approx \frac{g\delta y_1}{\nu'}\left(1 - \frac{\gamma''}{\gamma'}\right) + \frac{1}{\chi}\sqrt{g\delta\left(1 - \frac{\gamma''}{\gamma'}\right)}\left(\ln\frac{\delta}{y_1} - 1\right) \quad (5.273B)$$

Finally, multiplying both sides of this expression by δ/ν' and replacing χ and y_1 with their respective values, we obtain

$$Re = \frac{G_1'}{g\mu'} \cong \sqrt{\frac{g\delta^3}{\gamma'^2}\left(1 - \frac{\gamma''}{\gamma'}\right)}$$

$$\times \left\{11.6 + 2.5\left[\ln\left(\frac{1}{11.6}\sqrt{\frac{g\delta^3}{\nu'^2}\left(1 - \frac{\gamma''}{\gamma'}\right)}\right) - 1\right]\right\} \quad (5.274)$$

Thus, from information on the liquid film's physical properties and mass rate G'_1, the thickness of a turbulent film flowing down a vertical wall can be estimated. For further information on this subject, the reader is referred to the references.

NOTATION

a = contraction loss coefficient, see Equation (5.170)
a = bending angle, see Equation (5.176)
A = amortization costs, \$/yr
A = coefficient in Equation (5.158A)
a,b = fraction of costs for maintenance and repairs
a,b = sides of rectangle, m
C = discharge coefficient
c = coefficient defined by Equation (5.178)
C_e = cost per unit of energy, \$/kW
c_ℓ = cost per unit length of piping, \$/m
c_1 = velocity of sound, m/s
C_L = discharge coefficient
C_p, C_v = specific heat at constant pressure or constant volume, respectively
d_{eq} = equivalent diameter, m
D = diameter, m
D_0 = optimum pipe diameter, m
e = roughness height, mm
E_v = eddy viscosity, cp
E = energy costs, \$/yr
E = energy term, J/kg
Eu = Euler number
F = area, m^2
$\hat{F}$ = friction losses, N
Fr = Froude number
g = gravitational acceleration, m/s^2
G = superficial mass velocity, kg/m^2-hr
g_c = conversion factor, 32.174 ft-lb$_m$/lb$_f$-s^2
h_{fr}, h_{1r} = friction and local resistance losses, m
h_ℓ = lost head, m or J/kg
H = velocity head, m
H_0 = Homochronicity number
i = enthalpy, J/mol
I = turbulene intensity parameter
j = fractional cost
k = space factor, see Equation (5.149)

k_1 = cost parameter defined by Equation (5.221)
k_2 = cost parameter defined by Equation (5.225)
k_e = head loss coefficient
k_i = yearly expenses for amortization and maintenance, \$/yr
k_p = yearly production and operating expenses, \$/yr
ℓ = length, m
$\bar{L}$ = average work per unit area due to gravity, J/m²
L = length, m
L_e = equivalent length, m
L_T = turbulence scale, defined by Equation (5.37)
M = mass rate, kg/s
M' = viscosity scaling factor, μ/μ_0
m = parameter in Equation (5.158A)
n = exponent in Equation (5.209)
n = function defined in Equation (5.183)
N' = number of tube rows
N = pump power, hp
$\tilde{N}$ = energy cost for fluid transport, \$/hr
p = wetted perimeter, m
p' = density scaling factor, ϱ/ϱ_0
$\wp$ = pressure scaling factor, p/p_0
P_{cr} = critical pressure, Pa
P, p = pressure, Pa
q = exponent in Equation (5.158A)
$\dot{Q}$ = heat, J
Q = volumetric flow, cfs or m³/s
r = pipe radius or distance from wall, m
r_h = hydraulic radius, m
R = tube radius, m
Re = Reynolds number
$R(y)$ = correlation coefficient
s = function defined in Table 5.13
S = surface area or hole area, m²
t = time, s
t' = tube separation, mm
T = temperature, °C or °K
u = variable
V = volumetric flow rate, m³/hr
v = velocity, m/s
$w_{x,y,z}$ = velocity components, m/s
W = mass flow rate, kg/s
$\tilde{W}$ = average velocity scaling factor
W' = mechanical work, J

$\overline{W}$ = mean linear velocity, m/hr
W_g = weight rate defining flow regime for optimum pipe diameter, kg/s
$\hat{W}_s, \hat{W}_p$ = shaft power or mechanical power, ft-lb$_f$/lb$_m$ or kJ
$W_{x,y,z}$ = velocity scaling factors
x = parameter in Equation (5.209)
X = empirical parameter in Equation (5.176)
X', Y', Z' = distance scaling factors
y = distance from wall, m
Y = yearly operating time, hr/yr
Z, z = leveling height, m

Greek Symbols

α = correction coefficient for laminar or turbulent flow
β = diameter ratio or angle, °
Γ = mass rate per unit perimeter, kg/m-s
$\tilde{\Gamma}$ = length to diameter ratio
γ = specific weight of fluid
∇^2 = Laplacian operator, refer to Equation (5.56)
δ = film thickness, mm
$\hat{\delta}$ = velocity dampening parameter defined by Equation (5.189B)
ϵ = relative roughness, e/D
ϵ_M = eddy diffusivity of momentum (E_ν/ϱ), mm^2/s
ϵ_j = jet constriction coefficient
ζ = friction resistance coefficient, see Equation (5.154)
η = pump efficiency
θ = velocity component for compressible fluid, m/s; also transformation to cylindrical coordinates
$\varkappa$ = ratio of specific heats, C_p/C_v
λ = friction factor
λ_{is} = friction factor at specified temperature
μ = viscosity, cp
ν = kinematic viscosity, mm^2/s; or specific volume as noted, m^3/kg
ν' = volume, m^3; or specific volume, m^3/kg
ϱ = density, kg/m^3
σ = surface tension, N/m
τ = shear stress, lb$_f$/in^2 or Pa
τ_N, τ_T = viscous and turbulent stresses, respectively, lb$_f$/in^2 or Pa
T' = shear scaling factor, τ/τ_0
T = friction force, N
φ = velocity coefficient
ϕ = velocity factor

χ = constant defining turbulence structure, see Equation (5.265)
ψ = correction coefficient for nonisothermal friction factors or Weisbach equation discharge coefficient, see Equation (5.171)
ψ' = deviation coefficient, see Equation (5.253)

REFERENCES

BEIJ, K. H. "Pressure Losses of Fluid Flow in 90 Degree Pipe Bends," *J. Res. Nat. Bur. Standards* XXI (1938).

BENNETT, C. O. and J. F. Myers. *Momentum, Heat and Mass Transfer.* New York:McGraw-Hill Book Co. (1964).

BERGELIN, O.P. *Chem. Eng.*, 56(5):104–106 (1949).

BIRD, R. B., W. E. Stewart and E. N. Lightfoot. *Transport Phenomena.* New York:John Wiley & Sons, Inc. (1960).

BOUCHER, D. F. *Chem. Eng. Prog.*, 44(10):527,601 (1947).

BRODKEY, R. S. *The Phenomena of Fluid Motions.* Reading, MA:Addison-Wesley Publishing Co. (1967).

BROWN, G. et al. *Unit Operations.* New York:John Wiley & Sons, Inc. (1950).

CAMBEL, A. B. and B. H. Jennings. *Gas Dynamics.* New York:McGraw-Hill Book Co. (1958).

CHEREMISINOFF, N. P. *Applied Fluid Flow Measurement: Fundamentals and Technology.* New York:Marcel Dekker, Inc. (1979).

CHEREMISINOFF, N. P. *Fluid Flow: Pumps, Pipes and Channels.* Ann Arbor, MI:Ann Arbor Science Publishers (1981a).

CHEREMISINOFF, N. P. *Process Level Instrumentation and Control.* New York:Marcel Dekker, Inc. (1981b).

CHILTON, T. H. *Trans. Am. Inst. Chem. Eng.*, 29:161 (1938).

COOPER, C. M., T. D. Drew and W. H. McAdams. *Ind. Eng. Chem.*, 26:428–431 (1934).

CURLE, N. and H. J. Davies. *Modern Fluid Dynamics, Vol. 1: Incompressible Flow.* New York: D. Van Nostrand Co. (1968).

DAVIES, J. T. *Turbulence Phenomena.* New York:Academic Press, Inc. (1972).

DODGE, L. "How to Compute and Combine Fluid Flow Resistances in Components," *Hydraul. Pneum.*, 21(9):118–121 (1968).

FOLSOM, R. G. *Trans. Am. Soc. Mech. Eng.* 78:1447–1460 (1956).

FRIEDMAN, S. G. and C. O. Miller. *Ind. Eng. Chem.* 33:885–891 (1941).

HODSON, J. L. *Trans. ASME*, 51:303 (1939).

HOOPER, C. M. *Trans. Am. Soc. Mech. Eng.*, 72:1009–1110 (1950).

JACKSON, M. L. *Am. Inst. Chem. Eng. J.*, 1:231–240 (1955).

JAKOB, M. *Trans. ASME*, 60:384 (1938).

JESCHKE, J. *Tech. Mech. V. D. I.*, 69 (1925).

KAPITSA, L. P. "Volnovoye techenie Tonkikh Sloyev Vyazkoy Zhidkosti (Wave Flow of Thin Layers of Viscous Liquid), *Zhetf (J. Exp. Theoret. Phys.)*, No. 1 (1948).

KING, H. W. and E. F. Brater. *Handbook of Hydraulics.* New York:McGraw-Hill Book Co. (1963).

KING, R. C. and S. Crocker. *Piping Handbook*. New York:McGraw-Hill Book Co.) (1967).
KNUDSEN, J. G. and K. D. L. Katz. *Fluid Dynamics and Heat Transfer*. New York:McGraw-Hill Book Co. (1958).
KOVAKOV, P. K. *Izv. an SSSR, OTN*, 7:1029 (1949).
KRYLOV, A. V. *Dan SSSR*, 56(2):133 (1947).
KUTATELADZE, S. S. and M. A. Styrikovich. *Hydraulics of Gas-Liquid Systems*. Moscow:Wright Field Trans. F-TS-9814/V (1958).
LEVICH, V. G. *Physico-Chemical Hydrodynamics*. Englewood Cliffs, NJ:Prentice Hall, Inc. (1962).
LIEPMANN, H. W. and A. Roshko. *Elements of Gas Dynamics*. New York:John Wiley & Sons, Inc. (1957).
MINER, I. O. *Trans. Am. Soc. Mech. Eng.*, 78:475-479 (1956).
MOODY, L. F. *Trans. ASME*, 66:671 (1944).
MOTT, R. L. *Applied Fluid Mechanics*. Columbus, OH:Charles E. Mezzill Publishing Co. (1972).
NORMAND, C. E. *Ind. Eng. Chem.*, 40(5):783 (1948).
OTHMER, D. F. *Ind. Eng. Chem.*, 37(11):1112 (1945).
PAI, SHIH-I. *Viscous Flow Theory-Laminary Flow*. New York:D. Van Nostrand Co. (1956).
PERRY, J. H., ed. *Chemical Engineer's Handbook*, 3rd ed. New York:McGraw-Hill Book Co. (1950).
ROHSENOW, W. M. and J. P. Hartnett. *Handbook of Heat Transfer*. New York:McGraw-Hill Book Co. (1973).
ROUSE, H. *Elementary Mechanics of Fluid*. New York:John Wiley & Sons, Inc. (1948).
SHAPIRO, A. H. *The Dynamics and Thermodynamics of Compressible Fluid Flow*. New York:Roland Publishing Co. (1953).
SIEDEZ, E. N. *Ind. Eng. Chem.*, 28(12):1429 (1936).
SIMPSON, L. L. "Process Piping: Functional Design," *Chem. Eng.*, 76(8):167-181 (1969).
STOKES, R. L. *Ind. Eng. Chem.*, 38(6):622 (1946).
STOLL, A. W. *Trans. Am. Soc. Mech. Eng.*, 73:963-969 (1951).
STREETER, V. L. and E. B. Wylie. *Fluid Mechanics*. New York:McGraw-Hill Book Co. (1979).
WHITE, S. M. *Engineering*, 128:69 (1929a).
WHITE, S. M. *Proc. Roy. Soc. A.*, 123 (1929b).
WHITWELL, J. C. and D. S. Plumb. *Ind. Eng. Chem.*, 31(4):451 (1939).
WILSON, R. E. *Ind. Eng. Chem.*, 14(105) (1922).

BIBLIOGRAPHY

ASME Research Committee on Fluid Meters. *Their Theory and Application*. New York:The American Society of Mechanical Engineers (1959).
FALLAH, R., T. G. Hunter and A. W. Nash. *J. Soc. Chem. Inc.*, 53:369-379T (1934).
Flowmeter Computation Handbook. New York:American Society of Mechanical Engineers (1959).
Fluid Meters: Their Theory and Application, 5th ed. New York:American Society of Mechanical Engineers (1959).

GEANKOPOLIS, C. J. *Transport Processes and Unit Operations.* Boston, MA:Allyn and Bacon, Inc. (1978).
Greve-Bull., Purdue Univ., 12(5):32 (1928).
GUKHMAN, A. A. *Introduction to the Theory of Similarity.* New York:Academic Press, Inc. (1965).
LAPPLE, C. E. *Chem. Eng.*, 56(2):96–104 (1947).
SEDOV, L. E. *Similarity and Dimensional Methods in Mechanics.* New York:Academic Press, Inc. (1959).

INDEX

absolute manometer 96
absolute pressures 13
absolute scales 12
absorption tower 36
acceleration due to gravity 254
acceleration of gravity 13
acetaldehyde 37
acetic acid 79
adiabatic process 20
apparent viscosity 33
area moment of inertia 108
auto modeling region 64
average specific volume 164
average specific weight 164

bank of tubes 236
barometric pressure 13
base units 1
basic equation of hydrostatics 87
behavior of fluids 22
Bernoulli's equation 170, 171, 178, 182, 187, 235
Bernoulli's principle 189
Bingham model 31
bitumens 34
boundary conditions 61
Buckingham Pi Theorem 58

capillary tube 196
capillary waves 260
center of pressure 106
centipoise 76
channel configurations 231
characteristic parameters 45
classification of fluid behavior 25
coefficient of kinematic viscosity 76
coefficients of contraction 210
complexes 58
compressible fluids 73, 141

condensed phases 71
constitutive equations 22
constricted piping 234
continuity 186
continuity equation 135
contraction loss 223
contraction loss coefficient 210
critical discharge velocity 153
critical Reynolds number 125, 200
cylindrical nozzles 190

Darcy equation 219, 237
Darcy Weisbach equations 160, 214, 228, 229
deformation 23
density 12
density of gases 74
derived units 1, 5
differential equations for viscous fluids 139
differential manometer 100
differential pressure 94
dilatant fluids 30
dimensional analysis 43, 56
dimensionless number 57
dimensionless groups 49, 50, 51, 58, 60, 65, 69
discharge coefficient 183, 184, 188
discharge of gases 151
distribution systems 194
dynamic head 172, 173
dyne 7

eddies 133
eddy diffusivity of momentum 134
efflux from vessels 187
English system 7
enthalpy 16
equation isomorphism 62
equations 137
Euler number 50, 58, 147

271

Euler's differential equations 85
Euler's equations 86

Federal Register 1
Federman Buckingham's theorem 59
film flow 259
film interface 249
film thickness 251, 254, 260
first law of thermodynamics 15
flow behavior index 28
flow normal to tube banks 236
flow nozzle 182
flow resistance coefficient 157
flow through pipes 194
fluctuating velocities 132
fluid density 118
fluids with yield 31
friction 223, 263
friction coefficient 163
friction factor chart 205
friction loss 195, 225
friction resistance coefficient 196
frictional losses 174
frictional resistance 195
Froude number 58

gas constant 163
gas flow 258
gas flow through piping 156
gas installations 94
gases 22, 72
gauge pressures 13
geometric conditions 45
geometric configurations 44
geometric head 171
geometric scale factor 44
geometric similarity 63
gradual expansions 214
gravity 13

Hagen Poiseuille equation 145
head loss coefficient 219
heat capacity 17
homogeneous fluid 24
homogeneous slurries 26
horizontal pipe 195
hydraulic radii 231
hydraulic resistance 135, 173
hydraulic resistances in pipe flow 194
hydraulic seals 96
hydrodynamic head 171

hydrodynamic principles 194
hydrodynamic problems 71
hydrogen chloride offgas 36
hydromechanical processes 71
hydrostatic machines 111, 112
hydrostatic pressures 125
hydrostatic principles 90
hydrostatics 71, 85

ideal fluids 137
ideal gas 18, 19
ideal gas law 18, 152
ideal liquid 73
incompressibility 143
incompressible fluid 73, 178
incompressible homogeneous flow 171
industrial stoichiometry 6
interfacial disturbances 259
internal energy 15, 150
internal hydrodynamic problems 71
internal problems of hydrodynamics 115
International Organization of
  Standardization 1
international system of units 1
isobaric process 19
isothermal flow 156, 159
isotropic state 133

Kelvin scales 12
kinematic homochronicity 53
kinematic viscosity 76, 204, 221, 252, 257
kinetic energy 14, 148, 157, 209, 227, 235
kinetic processes 59
Kirpichev Gukhman's Theorem 61

Lagrangian derivative 116
laminar boundary layer 124
laminar flow 125, 199, 200, 249
laminar flow around tubes 237
laminar flow in coils 229
laminar friction factor 198
laminar regime 124
laminar sublayer 263
linear velocities 117
liquid 73
liquid film 249, 247
liquid pressure 81
liquid separator 97
liquid viscosity 258
local resistances 195
lost energy 174

Index 273

low molecular weight liquids 26
low pressures 94
low viscosity liquids 252

manometer fluid 93, 180
manometers 91, 92
manometric techniques 90
mass flow 119
mass flow rate 122
mass transfer 135
mass velocity 118, 122
Maxwell fluids 36
mechanical energy balance equation 213
methane 166
mixed friction 200
molten polymers 35
momentum balance 212

Navier Stokes equations 139, 141, 142, 147
negative pressures 94
Newton number 49, 50
Newton's law of viscosity 29, 75
Newton's theorem 47
Newtonian behavior 29
Newtonian fluids 26

one directional flow 143
operating expenses 242
optimum diameter 243
optimum pipe diameter 241
orifice 188
orifice meter 181, 182, 183

parabolic velocity distribution 127
parametric dimensionless numbers 58
Pascal's law 89, 111
physical properties 73
pipe components 241
pipe cross sectional area 227
pipe flow 115
piping 120
piping components 218
piping system 219
plastic fluids 31
Poiseuille's law 214, 232
polytropic process 21
potential energy 14, 89
pressure changes 72
pressure difference 228
pressure drop 162, 184
pressure forces 101

pressure measuring devices 98
pressurized air 96
principle of communicating vessels 90, 96
principle of continuity 190
process governing equations 52
pseudoplasticity 29

quicksand 31

Rankine scale 12
reduced state 77
regimes of flow 124
relative roughness 199
residence time 121
resistance coefficient 158
resistances for laminar flows 230
Reynolds number 50, 58, 203, 205, 221, 229, 242, 252
rheograms 27, 29
rheology 22
rheopectic 25
rheopectic fluids 34

self modeling zone 202
semicontinuous operations 39
shear forces 140
shear rates 25, 35
shear thickening fluids 29
shear thinning fluids 28
similarity conditions 45
similarity of boundary conditions 63
similarity of initial conditions 63
similarity of physical values 63
similarity theory 43, 44, 46
similarity transformation 54
single fluid flows 115
smooth friction 200
specific gravity 73, 105, 121, 174
specific heat 17
specific potential energy of pressure 89, 171
specific weight flow rate 162
static head 96, 173
steady state 115
steady state flow 143
Stokes law 127
Stokes number 51
stress relaxation 35
structural changes 25
substantial derivative 116
sudden expansion 208, 209
superficial water velocity 227

274 Index

supersonic conditions 154
supplementary units 1
surface tension 82, 83

tearing force 110
temperature 12
thermodynamics 16, 151
thixotropic 25
thixotropic fluids 32
thixotropy 33
three dimensional flow 140
time averaged velocity 131
time dependent fluids 25, 32
time independent fluids 26
time similarity 53, 63
Torricelli's theorem 189
total cost per unit length 241
total energy balance 148
total head loss 219
transformation techniques 145
transient flow 115, 116
transition region 125
true pressure 94
turbulence 134
turbulent core 263
turbulent film flow 124, 262
turbulent flow 124, 131, 132, 133
turbulent motion 202
turbulent regime 124, 199
turbulent velocity profile 135
two fluid manometer 100

uniform circular pipes 202
unit operations 37, 62
unsteady state 115

valves 241
variable head meters 178
varying cross sections 208
velocity dampening 233
velocity gradient 75
velocity profiles 233
ventilation system 123
venturi meter 181, 186
vertical film flow 247
viscoelastic 25
viscoelastic fluids 34, 36
viscosity 75, 76, 79, 143
viscous fluids 174
volumetric discharge 186
volumetric flow 122
volumetric flow rate 118, 204
volumetric rate 227

water gauges 90
wave motion 260
waves 259
wavy flow 257
Weissenberg effect 35
work 13

zone of quadratic resistance 200